It's time for change, pollution is killing the earth, smog is choking the air, it's time to clean up.

V's

"Clean Energy Sources Beyond 2020"

"Clean Energy Sources Beyond 2020"

Copyright Statement

The Need for Zero Cost Electricity

Zero Cost Electricity is the way forward due to the fact that it will drastically reduce:

- Air Pollution
- Spiralling Electricity costs
- Acid Rain

These reasons alone are enough to persuade Governments, Industries and individuals to encourage the use and further the development of Zero Cost Energy systems.

Alas! There will still be some industries and governments that will defer the development of green energy systems despite the serious detrimental affects of the resulting pollution on both nature and on mankind.

There is a growing need for a groundswell of popular support for alternative energy both at industrial and individual levels.

The public should have the freedom and support to use free energy systems in every application possible.

Some of the largest Nations are now becoming committed to Zero Cost Energy systems. Japan, China and India are already starting to reap the benefits of these clean air systems.

My challenge to you "Join the Green Revolution Now"and help save the planet.

Table Of Contents

Introduction

As most of you will be aware, air pollution is becoming a major problem worldwide. In nations like Japan, China and India air pollution has been so bad in some cities that people have resorted to wearing masks to protect their lungs from contaminants in the air.

Air pollution has been proven to cause a number of respiratory diseases including Bronchitus and Asthma.

If nothing changes these problems would only become more serious.

There is also the problem of acid rain that affects buildings by dissolving concrete. This acid also kills trees and vegetation.

These issues need to be fixed. The best solution is to encourage the use of "Zero Cost Electricity". All the zero cost systems that I have seen so far are also zero pollution systems, which means that they are an almost perfect solution to the problem of air pollution.

Common sense dictates that we need to inspire individuals, corporations and governments worldwide to adopt policies to encourage the use of"Free Energy Systems".

This is for the greater good of all humanity.

The purpose of this publication is not only to heighten awareness of the need for "Free Energy"but to act as a guide to showcase the main types free energy systems now available using conventional technologies. Most people know of Solar Panels, but there are other systems that maybe more suitable to your needs. It is my aim to describe these systems to you so that you can make a properly informed choice. These systems will be described, so that you can compare them. I will give you my evaluations to help you in your decision making.

This book is about clean energy systems not just for households but also for the electrical power industry.

The old ways of generating electricity are no longer sustainable.

These old power stations pour out pollution into the atmosphere through their industrial size chimneys.

Its not just industry and powerstations that are causing air pollution. It is motor vehicles running on fossil fuels that are belching out pollutants despite catalytic converters. Motor vehicles add an enormous amount of carbon and other pollutants into the atmosphere every day, of every year.

We must change our ways to correct the problems caused by pollution.

Chapter 1

Outlining the Problem

The term 'Greenhouse Gases' is derived from the way greenhouses work to keep heat inside, keeping plants warm in cold climates. The heat is trapped by the glass.

In much the same way certain gases emitted into the atmosphere form a layer that traps heat around the earth. Without the layer of greenhouse gases this heat would have escaped out into space, acting as a heat exchange mechanism for the earth.

The 'Greenhouse Effect' is a naturally occurring property of our atmosphere which allows life to flourish, for without it the earth would be too cold for life to exist.

The Earth's atmosphere naturally contains greenhouse gases, but they have existed within a natural balance.
The most common of these are; Water vapour; Nitrogen 78%, Oxygen 21%; and Argon 0.9%, with the trace gases CO_2, Nitrous oxide, methane and ozone accounting for one tenth of 1% in total. There are also minor amounts of Chlorofluorocarbons [CFCs] and Hydrofluorocarbons [HFC's].

The problem with greenhouse gases is that man has caused rapid gas emissions that are excessive and out of balance with the natural order.

The earth also has its own shield against extreme Ultraviolet rays from the sun. This shield was called the Ozone layer which has also been affected by manmade gases which have depleted this ozone layer and caused holes that can allow more heat to reach the earth than is normal. This heat is in the form of UV rays which can be very damaging.

Together these two extra heat sources have caused an effect called "Global Warming". Many of the effects of global warming have mentioned previously, but it has scientists and politicians and concerned citizens worried.

The effects can be catastrophic and long lasting, much of which has been covered in the introduction.

This increase in atmospheric gases can be traced back to the Industrial revolution. From the time of the Industrial Revolution to the modern day, the activities of mankind through land clearing, the burning of fossil fuels including coal, oil and natural gas are increasing concentrations of greenhouse gases in the atmosphere.

The threat of global warming is one that must be taken seriously. Action must be taken to combat global warming at its source. Mankind must be smarter about the way that we generate electricity, about the way that we fuel our transport vehicles.
Industries of all types need to actively find ways to reduce or eliminate gas emissions at the source.

Part of the answer to this problem is to put electric vehicles on the roads of the world. Unfortunately some nations of the world fall shamefully behind in this goal. These nations, Australia, the USA and probably Britain, are behind as their governments are growing fat from oil taxes imposed at the petrol bowsers. This needs to change. Having hybrid vehicles on the roads is pointless when vehicles can now be powered entirely by electricity.

It's time to get serious about electric cars. Around the world today there are nearly as many cars as there are people. The Petrol and Diesel motors pour out noxious gases into the atmosphere despite the catalytic converters. Hybrid cars are built to appease the big oil companies. In order to tackle air pollution there needs to be a push for the production of vehicles powered entirely by electricity.
This is an easily achievable objective today, back in the 1920 there were many electric vehicles being used on the roads. Electric cars are not some novel new craze, now is the time to bring back the electric car.

The green movement is about saving the planet and leaving it to our children in a better state than we found it in.

It is not just cars but modern industry and especially electricity generating plants have contributed very heavily to our polluted atmosphere.

This modern world in which we live has become reliant upon electricity, for all kinds of devices from Toasters and Heating and Cooling, and a myriad of other electrical products.

Not all things run on mains electricity, some are battery powered, and there are now many different types of batteries.

Unfortunately, in the past we have had little option but to obtain our mains electricity from large electrical supply corporations. These suppliers were mostly generated their electricity using coal fired stations. In recent years deregulation of electricity suppliers has occurred in many countries. These electricity suppliers are now unregulated and can charge whatever fees they choose whether this is a fair price or not. In recent years, we have seen electricity prices in most countries spiral to uncomfortably high rates.

Since deregulation electricity prices in Australia have been the highest in the world. This shameful price gouging is an insult to the Australian people and a black mark on the Australian Electricity Industry.

The Australian government promised to bring electricity prices down but we are all still waiting without seeing any significant results, which is an inditement against the Australian Goverment, who should learn to deliver on their promises, not just give false hope.

It is not just in Australia but many of the electrical generating plants around the world were coal fired, emitting clouds of carbon gases into the atmosphere, many of them are still operating. The worlds combined emissions from all sources including vehicles is estimated at 37 gigatonnes of CO_2 into the air.

When the Carbon dioxide in clouds mixes with the hydrogen ion it produces Carbonic acid which falls as rain, damaging trees and etching concrete due to its acidic nature. Acid rain has caused much damage to the black forrest region of Germany.

Solar and other sources will help solve the acid rain problem along with climate change, by restricting the amount of CO_2 being fed into the atmosphere by the endeavours of mankind.

The emergence of Solar panels into the marketplace some 30 to 40 years ago has seen a lower dependence of many homes on industrial electricity supplies.

Switching to Solar is set to reduce Carbon Dioxide emissions by over 37 million metric tons per year.

Clean air has become a very big issue worldwide not just because of health issues but due to the fact that people believe air pollution is causing global warming. The "New York Times" recently reported that India's air pollution was causing the deaths of 1.1million people per year only surpassed by China's death toll from air pollution.

Needless to say, these two Nations are now leading the race towards cleaner air quality. The Indian government has recently embraced the shift from coal to the cleaner renewable energy sources, chiefly solar. The worlds biggest coal company 'Coal India' has recently shut down mining operations at 37 coal mines in India.
India's Energy Minister was quoted as saying "Our focus is now on renewable energy. The Indian government is encouraging the use and development of solar power.

The energy revolution is being lead by India in the wake of plummeting solar prices, and the government incentives are making renewable energy sources more attractive.

Changes in the design of solar panels now allow for efficient running of solar power in even the coldest climates. The result is for all or most of the year there are significant savings on electricity costs. A large majority of solar users claim to be able to supply all their own household electricity at zero cost to themselves. Some of these solar households even produce excess electricity that they sell back to the grid.

Solar cells have become increasingly more Climate resilient over the last 10 years particularly in Germany which recieves half the sunshine of the sunniest cities of the USA. Germany now, also has the most successful solar power system in the world.

Solar is not the only option for free electricity, not the cheapest to install or the most efficient, more to come.
The scope of this book includes clean energy systems that can be used by the householder but also includes clean energy systems that can be used by industry.

One of the main objectives is to create a world in which we have healthy breathable air. This will benefit all mankind.
One way that Big Oil can help this clean air movement is to become part of it. There is an abundance of money to be made in the green revolution. Even the people that work for big

oil will benefit from clean air.

Slowing down global warming will also benefit everyone.
Global warming will affect every single one of us. Is global warming really happening?
Recently scientists took reading of the earth crust temperature via satellite and the results confirmed that the earth is indeed warming up.

So then, what can you expect from global warming?
Here are some of the affects of global warming:
An increase in average temperatures worldwide
Plants dying due to heat
Melting ice caps, causing global sea levels to rise at a rate of 3.2mm per year. This may increase as more ice melts from polar regions and mountain glaciers.
Changes in weather patterns, increased severity of hurricanes and storms, severe hail storms, longer droughts.
Water scarcities in some areas
Warmer sea waters causing the death of coral reefs.

If humanity can do anything to slow down or stop global warming, it needs to be done now, before it becomes worse.

Quote by Intergovernmental Panel on Climate Change
"Taken as a whole, the range of published evidence indicates that the net cost of climate change are likely to be significant and to increase over time." [Global Climate Change, Nasa]

Climate change could be happening at a rate 10x faster than was originally calculated. Island nations are at high risk from erosion and rising waters and more powerful storms.

Only time will reveal if the above statements are even close to being accurate, often we worry about things that don't even eventuate.

Chapter 2

Renewable Energy Sources

Looking for Solutions

The term 'Clean Energy Sources' also encompases the terms "Sustainable energy", Renewable energy, Green energy, Eco-friendly energy, and Green Technology.

Renewable Energy Sources, often shortened to simply 'Renewables' is a term for energy sources that are easy to replenish and has come to also include the notion of clean energy sources.

Renewable energy is defined as energy derived from sources that do not deplete or can be restored within a human lifespan. The most common of these include; Geothermal, Biomass, Hydropower, Solar Thermal and Solar Electric, Tidal and Wind power sources.
But are all of these truly clean energy sources?

In many nations around the world renewable energy sources account for approx 20% of the total energy production by those nations and this trend towards renewable-energy is set to continue. Iceland and Norway already produce all their electricity through renewable sources. Other nations are following suite with projections of reaching 100% renewable energy in the future.

According to Wikipedia, 47 nations around the globe have already reached a target electricity production of 50% from renewable sources.

In the early years of renewable energy production, the costs were very high for renewables compared to fossil fuels. As these forms of energy producing methods have received funding and research and development, the costs associated with producing renewable energy has dropped dramatically and now many of them are cheaper to produce than by using fossil fuels. Costs now tend to favour renewable energy sources, which will drive the production and development of fossil fuels even further.

There are many large scale renewable energy projects, renewable energy technologies can also be successfully adapted to rural and remote areas, even in developing countries.

As nations develop, it also drives the development of electrification, by renewables which is now cheaper than fossil fuels and is a major advantage to developing nations.
The lack of pollution, especially air pollution is a benefit to any nation, being a much healthier option for its people.

Geothermal energy is an option for a few nations but it is restricted to geographical locations that have access to thermal heat from the earths core. New Zealand is probably a good candidate for geothermal energy due to its access to boiling mud pits, which are a product of geothermal energy.
Geothermal energy is outside the scope of this book due to the limited number of locations suitable to the extraction of this energy source.

Biomass is considered a renewable energy source, it involves converting Garbage; Wood; Crops; Landfill gas and Alcohol fuels into electricity. The main method requires the biomass to be burned, thus releasing CO_2 into the atmosphere and will not be considered a clean energy source for the purpose of this book. I am not against this type of energy being used but it should be restricted to small scale use, not industrial.

Hydroelectricity schemes should also be discounted unless they can be designed to be less harmful to the environment.
The Snowy Mountain scheme whilst a success for generating electricity was an enormous loss for the environment, causing irreversible damage to rivers in the Snowy Mountains area.

Fish that once populated these rivers have never fully recovered from the changes to their natural habitats caused by daming rivers and changing natural water flows.
This was probably an oversight by the developers but it is one that should not be repeated.

There have recently been talks of a Snowy Mountains Hydroelectric Scheme Two. In the aftermath of the first Snowy Mountains Scheme, a second would most likely kill more river systems and devastate more natural wildlife.
How many river systems need to die when we already have viable alternatives, such as Solar

Thermal, Solar Electric, Wind power, both Turbines and Chimneys, Tidal, and now even two different sources of clean Nuclear power sources at our command.

Another system that has aroused a lot of attention recently is the new regenerative coal fired plants that are claimed to be much cleaner than the older designs that spewed CO_2 into the air in bulk unceasing quantities.

Whilst most of the world is turning its back on coal fired plants as a means to produce electricity, one nation has been holding firm, that nation is Japan.

New Technology has been developed in the past 10-15 years by the coal industry which is vying to remain contemporary in the face of growing concern over the CO_2 being dumped into the atmosphere.

These new technologies are called UltraSupercritical and Circulated Fluid Bed [CFB] Technologies which improve the combustion side and the emission control side of the combustion of coal.

Even with these new modifications, coaled fired plants still release a certain amount of CO_2 into the air. These new plants burn less coal and produce less greenhouse gases.
The question that remains is:

"Are Coal fired electric generating plants really necessary when electricity from other clean energy sources are becoming more widely available and at costs that are now very competitive or better than fossil fuel sources."

For now I cannot recommend coal fired power plants as clean energy sources. Unless it it the new design called regenerative coal fired power stations that filter out most of the carbon rather than expelling it into the atmosphere.

Chapter 3

Solar Electricity – Part 1 Solar Panels

Photo's: A few Solar Powered Homes in Southern Adelaide on a cloudy day the 2nd May 2019, these were snapped by the author.

Solar panels are multiplying everywhere, with most new houses in Australia opting to include the cost in their housing loans.

Solar panels are made up of Solar cells grouped together in a panel type framework. The most common solar cells are called photovoltaic cells or PV cells. There are two basic types of PV cells, these are made from semiconductor materials, chiefly various forms of silicon. The other modules have been manufactured as thin films, and are called thin-film PV. Each type of solar cell has different properties, cost and performance characteristics. Certain thin film PV units can be applied to windows and glass panes, and are often seen on skyscrapers and high rise buildings.

In Australia 2019 could be the optimum time to be installing solar energy with the government rebates that are available for this purpose. These rebates will not last forever but can save you upto $6000 on your solar system. For more information, go to: solartips.com.au

Solar panels are a popular and growing trend among homeowners who want to reduce or totally eliminate their electricity costs. Installing Solar panels itself is not without cost. There are the obvious costs of purchasing the panels and then the cost of installation, including connecting them to your homes power supply.

Excess power produced during daylight hours is put into a storage system, normally it will be a batteryback-up system. The older systems used either lead acid batteries or deep cycle batteries connected in a network. These systems are an eyesore unless they are mounted in frames and hidden in a cupboard.

The newer batteryback-up systems tend to be lithium ion batteries which are installed into a box type casing and mounted on a wall. The lithium ion battery option is smaller and lighter than the old battery types.

An energy audit can be done to determine how much electricity you need to run your house. However, it is said that most houses can run on 10Kw. If your house is larger than the average size home than you will need a slightly larger capacity unit.

Current Solar panel electrical systems look good although they are still only about 20% to maximum of 25% efficient. In the recent past they would not operate well during cloudy winters or in areas where there is a lot of snowfall. The newer panels will operate and be highly functional in these cold climates as conductivity now increases at colder temperatures.

Modern solar panels have vastly improved from the 1960's and are now suitable for for most climate conditions. However if you have a dusting of snow you may want to clean the snow off the panels for peak performance.

Solar Panels are popping up everywhere these day, from caravans, farm buildings, barns, street lighting, phone booths to high rise buildings. It seems that everyone wants to save money on electrical costs.

Solar energy output peaks when the sun is at its highest point in the sky, about mid-day. At this time solar panels are at their peak performance offsetting expensive electricity when demand is the highest. Homeowners will need about 24 units to reduce or eliminate electricity bills.

Photo Courtesy Pixabay.com

Evaluation

Despite the current design of Solar Panels they are definitely suitable for use by householders to reduce or eliminate electricity bills. They are well suited to being fitted to the rooftops of houses and garages and have a pleasing look to the eye.

Warning

The householder needs to be aware that there are a lot of shonky operators in the Solar energy field, Shonky installers, and shonky solar retailers. No one wants overpriced installation or to be ripped off by operators offering lease payment options which can put the householder into debt for upto 30 years or more. Its bad enough that standard solar panels are priced to be a rip-off. My advice is if you intend to have a Solar system installed make sure you go through a reputable dealer who uses professional installers who charge reasonable rates. Know all the costs before proceeding and make sure that your bills will be smaller not higher. The dealer should be able to advise you on the aproximate cost of your electricity after installation.

Cost

The cost of Solar Panels will depend on the brand, the number of solar panels fitted and will include installation costs and the inverter. The number of Solar Panels that you will need to eliminate your power bill will vary according to your household needs but on average at least 24 panels will be needed. In Australia systems vary from roughly $2500 AUD to $6000 just for the solar panels.

When you get a quote for Solar Panels make sure it includes all these elements. The Panels, The Inverter, Battery backup and installation costs.

The Tesla Lithium Ion Battery Backup system will cost from $6000 AUD. In Australia, the government is offering rebates on Solar systems, so you might want to investigate these rebates.
In Australia for tips on Solar go to: lp1.solartips.com.au

The Future of Solar Panels

Current solar panels work well but are considered inefficient as the only convert the white spectrum of light into electricity. There is under development a new breed of Solar Cell that is said to be 44% efficient by converting the whole light spectrum into electricity. This is roughly double the efficiency of the current cells.

Solar panels that are 44% efficient would literally mean that you would only need half the number of solar panel to power your home.

Quote:
"During the past 40 years, solar prices have dropped 250-fold. And as these costs plummet, solar panel capacity continues to grow exponentially."
[Singularity Hub: https://singularityhub.com/2019/05/17/5-coming-breakthroughs-in-energy-and-transportation/].

Solar Panels have an expected lifespan of approximately 30 years, generally speaking.

Cells with 50% Efficiency
There is a new solar cell being developed, the cell is designed to absorb both white light and the longer wavelenth spectrum of light which could raise the performance of the cell to 50% efficiency.
Solar cell design with over 50% energy-conversion efficiency
https://phys.org/news/2017-04-solar-cell-energy-conversion-efficiency.html
APRIL 24, 2017 Kobi University

Thin Film Solar Cells
A more recent development than the semiconductor PV solar cells is the thin film solar cells which are printed onto thin plastic like sheets. They are about 20% efficient whereas the semiconductor type are now about 30% efficient.

Advantages of thin film solar are that they are very lightweight and can stick to almost any surface, making them ideal for high rise buildings. The guys at MIT under Vladimir Bulovic have been working on printed solar cells and inks that absorb light in the Infrared and UV ranges, this light is invisible to the human eye, making a totally clear solar cell possible, which can be printed on glass or on plastic sheets.
These solar cells are very cool simply because of the fact that they are transparent, allowing them to be used on glass instrument panels and provide power for the device. These cell also work in artificial light and in moonlight if its bright enough.

Some very cool gadgets could be made using these thin film cells.

Much of this research was done back in the year 2011, which begs the question.
Why has nobody financed this supercool breakthrough technology. This is technology that we need out on the streets NOW!

Investors, financiers, watch the youtube video below, and tell me that this technology is not worth backing.

Rethinking Solar Energy | Vladimir Bulovic | TEDxBeaconStreet
https://www.youtube.com/watch?v=WoS7PGFjwDQ

Printed Circuit Solar Cells
Solar Cells can be printed onto media like paper or plastic. There are a number of ways of doing this, the main ones are:
Screen Printing
- Using an ink printer, like an ink jet.
  Hand printing
  Vladimir Bulovic uses transparent inks that absorb both UV and Infrared. Do these inks also convert the trapped light into electric current.

It is now possible to print solar cells directly onto paper or plastic using an inkjet printer, the materials maybe different to standard cells but the working principles remain the same.
More information is available at:

Momentum Energy
Printed Solar Cells
https://www.momentumenergy.com.au/habitat/renewable-energy/solar-cells-printable

Printed Solar cells make a good D.I.Y. Project.

Evaluation of Solar Panel Systems

Efficiency: Current Solar Panels are very low on efficiency at between 20 -25% Maximum.

Cost: In Australia average price range is between $2500 -$4,500 depending on the number of solar panels that you have fitted. You will need about 24 panels to cover all your electricity needs.

Usage: High rise buildings, Farm buildings, Houses, Street lighting, Phone boxes, some cars, solar panels in their wide range of sizes are being employed in a large number of places.

Appeal: They look good so have a high appeal.

Durability: Good durability, but they may not be in good shape after a hail storm.

Life Expectancy: You are looking at 30 years according to SEIA. Most manufacturers will have a guarentee on their products, upto about 10yrs. It is always a good question to ask when purchasing Solar Panels.

Limitations: In very cloudy climates, or where there is lots of snowfall they may prove to be dissappointing in their performance.

Suitability: Extremely well suite to householder power supplies. Most average households require about 10.5Kw per hour, to run. You can use this figure as a rough guide, but an energy audit will give you a more accurate figure.

Backup: Most backup system are by battery. Lithium ion battery back-up systems are becoming very popular, although the old lead acid car batteries will be cheaper especially if you use second hand batteries.

General Comments: You really need to look into the cost of your Solar system as it can work out much more expensive to purchase and install than you may expect. Make sure your quote includes all the costs. Prices will keep dropping over time.

Future: Keep your eyes open for a new generation of Solar Panels being developed and tested, they are said to be 44% efficient by using the whole spectrum of light not just the white spectrum. Recently in Abu Dabai in the UAE there has been a breakthrough in running costs of their solar which has brought the cost of production down to 6c per kilowatt hour. A rate that is cheaper than producing the same amount of electricity with fossil fuels. Solar technology is now competitive with fossil fuels, and are becoming cheaper than the fossil fuels.

Solar Electricity – Part 2 Battery Backup

Standard Solar Panels only produce usable electricity during daylight hours. Therefore if you want to use electricity at night you can choose to use power from the grid which you will have to pay, or you can have your own batteryback-up system. The vast majority of people opt for a batteryback-up system so that their electricity usage remains free of cost.

Before Lithium Ion batteries became popular and abundant, the batteries used for storage were either Lead Acid that are used in cars or deep cycle batteries. These big old batteries were suitable to the task but were bulky and tended to look unsightly unless the were surrounded by a cabinet. Most of these large battery type networks were not really portable, so your options on where to place the was rather limited.

Due to a number of factors Lithium Ion batteries have become the battery of choice for most Solar Electric Panels as a backup storage for excess electricity in the last few years. These batteries have excellent storage capacity for their small weight and size.

As back up units they are being produced enclosed in small cabinets and are usually wall mounted on either the side of the house or on a garage wall. They are commercially produced now, by a number of different manufacturers, the most well known is the Tesla brand by Elon Musk called the 'Powerwall'.

Sonnen is another famous lithium ion battery manufacturer for solar panels.

Government Incentives or Rebates
>   Australian Government Incentives: lp1.solartips.com.au
>   South Australian Government: Solar Battery Subsidy Scheme
>   For Average household usage US residents check online with the Dept. of Energy.
[These are a ballpark estimate only].
>   USA: Search out the facts of solar for your home by speaking to a Solar Power
Advocate such as Understand Solar to get a proper Estimate for your home [This will be a more accurate quote].

The Future of Lithium Ion Batteries

Elon Musk's company Tesla is currently manufacturing plant is in Freemont California the size of 107 football fields. Tesla will also be building a Mega Factory for Lithium ion batteries in China. This can only be good for the future of Lithium Ion Batteries.

"Where are Tesla's factories based? Including Elon Musk's gigantic gigafactory vision in China." By Felix Todd
https://www.compelo.com/tesla-factories-elon-musk-gigafactory

A Better Home Solution

Lithium Ion Batteries may be capable of home energy storage and have the appeal of looking good and are small enough to be attached to a wall, however they only perform the one task, energy storage.

Some of the new Water Batteries are able to apply themselves to two tasks, a double whammy. They can store energy and act as a hot water system at the same time, killing two birds with one stone. This dual function is an enormous advantage to the householder both function wise and price wise. If these systems come at a reasonable cost, they would be much cheaper than Lithium Ion storage systems.

Evaluation of Lead Acid Battery back-up

Efficiency: In user terms these batteries are efficient and a good, cheap alternative to Lithium Ion but to the engineer they probably rate below 50% efficiency.

Cost: A much cheaper alternative to Lithium Ion systems, old lead acid batteries can be used and reconditioned if needed. These batteries are connected as a network.

Usage: Lead Acid batteries have long proven track history

Appeal: Low appeal due to the bulky size and the large connectors.

Durability: Normally guarenteed for 1-2 years, but we all know that they can last a lot longer than that. It is also possible to recondition them restoring them to near perfect working condition.

Life Expectancy: Should be able to expect 20 years or more but only if you have them reconditioned when needed.

Limitations: Space could be a limiting factor. They operate well in most weather conditions.

Suitability: Extremely apt for the task of energy storage.

Fitting & Installation: Will require an electrician to install to household power.

General Comments: These old batteries have proven themselves time and again, and remain the cheapest energy storage batteries in the world today.

Future: Lithium ion are gaining in popularity, Silicon Ion may yet eclipse both in the future.

Evaluation of Powerwall Systems

Efficiency: Very good. Relies on Lithium ion batteries.
Cost: Elon Musk's Powerwall is the most well known brand of Lithium Ion Batteryback-up for Solar Panels. The cost will set you back over $6,000 AUD and in the authors opinion this price is way too high for the average householder, but if you are wealthy, go for it. If you are paying for the Solar Panels and the powerwall and the inverter all at the same time this could be a costly exercise. Other manufacturers will have their own version of the powerwall. There is plenty of competition in the market for the sale and installation of solar panels. Shop around and Get good advice.
Usage: Powerwall is mostly just for houses or buildings that use similar amount of power.
Appeal: They are pleasing to the eye.
Durability: They are well built and attached to a wall, so they should last well.
Life Expectancy: This will depend on how many times the batteries can be cycled
Limitations: There maybe limits on operating in some environmental temperatures.
Suitability: Very suitable to the task at hand which is energy storage, but expensive.
Fitting & Installation: Will require an electrician to install to household power.
General Comments: The tesla Powerwall will be too expensive for most people, unless there is a drastic price drop. There are other brands on the market, you should check and compare. Don't forget you could opt for Lead Acid batteries, which are bulkier, heavier, than Lithium Ion. D.I.Y maybe an option for you, but remember, the actual connection will probably require an electrician by law.
Future: New developments are constantly on the rise. A similar battery is being developed based on the Silicon ion, said to be much less toxic than the lithium and expectations of a much greater energy yield.
Rebates or Incentices: See if you are eligible for a government rebate. Some governments are offering rebates to encourage the use of solar power.

Over-all comment: Still way too expensive, a rip-off.

D.I.Y Solar Panels

Sources:

It is also possible to make your own solar panels.

Some solar cells use copper circuits which have PV coatings applied over them.

Research "Printed Solar Cells", Screen Printing Solar-cells.

A new Solar Power Plant DIY - Make one with Kreosan
https://www.youtube.com/watch?v=OUICh3QdEx8

Home Made Solar Cells
https://www.fuellesspower.com/26_Solar_cell2.htm

D.I.Y Lithium Ion Batteryback-up [PowerWall]

Source: Jehu Garcia's DIY PowerWall V2.01

https://www.pcbway.com/project/shareproject/Jehu_s_DIY_PowerWall_PCB_V2_01.html

Watch these videos. https://youtu.be/9YwErplHps8. https://youtu.be/ LkLW9wBmlMs. 2nd - Order PCBs. 3th - Rest of Parts can be found

Standoffs - https://j35.us/M4-25Brass-standoff-ali

Fuses - https://j35.us/100SMDfuses-Ali

4cell Holders - https://j35.us/4cell-Holder-Ali

3cell Holders - https://j35.us/3cell-Holder-Ali

Ribbon Cable - https://j35.us/16pin-RibbonCable

IDC Connector - https://j35.us/16pin-IDC-Connector

IDC Socket - https://j35.us/2x8-RightAngle-IDC-socket

90 degree XT60 - https://j35.us/XT60PW-ali

Solar Electricity – Part 3 Solar Farms/Solar Parks

Solar farms are another eco-friendly electricity generation system that can be used to replace the old dirty technology, like coal burning power stations. In other words they are a clean electricity generating station. Utility companies can provide pollution free electricity to their clients by building large solar PV parks.

The early solar farms were not really economical in terms of production of electricity compared with fossil fuel burning stations.

Since around the early 70's there has been much more focus on green technologies, leading to more and more research by many different companies and with government incentives being available, it has spurred on the development of ever increasing advances in solar technologies both solar photovoltaics and solar thermal.

Solar PV farms tend to cover vast areas of land with a sea of solar panels, which are now becoming a serious contender in the electricity marketplace. The capacity of industry is now bringing the cost of solar down to all time lows, which is good for the supplier and good for the consumer.

Solar cell designs are continuously being improved due to the large number of global manufacturers who are churning out solar panels in enormous quantities as the marketplace is embracing the solar farm concept.

Over the period of about 50 years, the solar industry has been able to become competitive with fossil fuels, which was once thought to be just a pipe-dream, it is now a reality. However, the is even better news from an unusual source.

In the Arab city of Abu Dabai, it has been revealed that for the first time ever, a solar farm has produced electricity at cheaper rates than the fossil fuel burning power stations. It is now cheaper to produce solar power than dirty power.

Another thing that has plagued the solar industry has been the fact that solar only works during the daytime, but night time is when power usage peaks. This problem has now also been solved by connecting solar farms to huge batteries, thus at night, power can be drawn from the batteries allowing for a continuous supply both day and night.

Don Sadoway has perfected a Liquid Metal battery which is now ready for the market. It has been fully tested and is not prone to catching fire or explosion like some of the Lithium ion batteries have been in the past. This battery was purpose designed to operate alongside solar farms to store excess power generated during the daylight hours to be a source of electric power at night when there is the most demand for electrical power. Don has set up a company to manufacture the Liquid Metal Battery called Ambri. This is a big battery like the Tesla big Lithium ion battery.

For more information on Liquid Metal Battery, See:
      The battery that could make mass solar and wind power viable.
      Dispatch
      https://www.youtube.com/watch?v=ImqmMOkANgg

      Innovation in Stationary Electricity Storage: The Liquid Metal Battery.
      Don Sadoway
      https://www.youtube.com/watch?v=pDxegcZqx_8

Evaluation of Solar Farms/Solar Parks 2019

Efficiency: Solar cells are still inefficient but they produce free clean electricity.
Cost: Solar Panels and the connecting technology has fallen dramatically in recent years, and the Abu Dabai site is now producing electricity cheaper than the equivalent electricity produced by fossil fuels.
Usage:
Appeal: These Solar farms are not ugly
Durability: Solar Panels will probably need replacing after 20 years or less
Limitations: The inability of Solar Panels to store energy at night is a limiting factor. See General Comments.
Suitability: Extremely suitable for producing bulk electricity.
General Comments: Solar Farms could benefit greatly by having an industrial size battery like the Liquid metal battery for night time use.
Future: Due to global warming issues, Solar farms have a bright future ahead of them.

Chapter 4

Solar Thermal Energy Sources

Solar Steam is another source of green energy which needs further research and development. There are now solar thermal plants all around the world. Two notable designs that are being used today as a source of energy are located in India and in Australia.

Solar steam can be used as a source of power for the following applications:
Cooking, both home cooking and cooking on a commercial scale.
Electricity generation
There are a number of different ways to generate electricity from solar steam. These include using a turbine like wind power, using solar heating towers, and using solar dishes, solar troughs.
These can be used for, Heating, Refrigeration/Cooling

Solar Cooking

Solar cooking has been around for many years now, although it has not as yet found widespread use. They can be used to heat, cook and even sterilise liquids and other foods. As with most solar thermal technologies solar cooking is achieved by concentrating heat from the sun. Sunlight is focused upon a smaller area, resulting in a concentration of the radiant heat, hot enough to grill, sear, or bake. You can cook almost anything with solar cookers.

Solar cooking does not need fuels and costs nothing to operate, thus they are free from air pollution and their use can reduce deforestation and land deterioration caused by gathering and using wood for fires.

Solar cookers come in a miriad of different design to suite all types of cooking. They all work on the same principle of concentrating sunlight by using highly reflective surfaces. The heat can then be trapped by using glass lids. The temperature of your solar cooker will be determined by the design and orientation to the sun. Temperatures can range from 65°C to 400°C. Solar furnaces can produce temperatures well over 1000°C, but this type of extreme heat is not needed for cooking.

Cookers come in different shaped with the most common being the dish shaped and then the box type ones, and parabolic trough shape not shown below.

There are even hybrid systems that can be used with solar panels, so they are both solar thermal and solar electric.

Cooking with Fresnel Lenses
You can use fresnel lenses for cooking, they come in a number of different types but the best type for cooking is the linear fresnel which focus the suns heat to a larger surface area than the spot fresnels. Fresnels could be used in additional to dish cookers, or you can find their focal spot and place your cooking vessel in the zone of the cooking spot. You should wear sunglasses when using fresnel lenses as the concentrated light can be very bright white light.

Fresnel lenses can also be used as heat sources for furnaces. They can focus the suns heat to 3002°F/1177°C.

Fresnel lenses can be teamed up with solar troughs to produce steam.

Another use for fresnel lenses is to heat water even to the point of producing steam. Steam can be used for cooking or it can be used to drive a steam engine. See:
    STIRLING ENGINE FRESNEL
    Lens on a Steek Solar powered Stirling Engine
    https://www.youtube.com/watch?v=CDCTANU8Tfk

Large Scale Cooking
In India, Solar Steam cooking has been taken to a whole new level, with solar steam now

able to cater for 10,000 meals at a time. There are a number of large solar cooking facilities in India today. They don't all use the same technology. In Auroville the solar kitchen utilises a stationary solar bowl acts as a large semi spherical reflector that has a reciever which tracks the sun resulting in steam for cooking. This cooker is also capable of drying and pasteurization. This solar kitchen is used to prepare 2,000 meals per day.

Many of the other solar kitchens in India use another solar concentrating technology called the Scheffler reflector. In 1986 it was developed by Wolfgang Scheffler and consists of a parabolic dish reflector which is able to track the sun because it uses single axis tracking. This device because of its superior concentrating abilities reaches temperatures in the range of 450-600°C, making the task of solar cooking easier. The Scheffler cooker built in Rajasthan India in 1999 is able to cook 35,000 meals daily.

Since this time there have been 2,000 Scheffler designs built throughout the world.

The CSIRO have recently developed a solar system that does 3 jobs in one. Solar heat is used to provide both space heating and cooling as well as hot water. Sounds like good value.
> Solar cooling for Australian homes
> https://blog.csiro.au/solar-cooling-for-australian-homes/
> By Chris Johnson, 19 July 2013

> Australian expertise in Solar air-conditioning – CSIRO
> https://www.youtube.com/watch?v=FXId9VS2Z6k

This new 3 in 1 system is very impressive, and by now it should have been commercialised and market ready. This design is definitely a big money saver.
Some of the devices used to help focus and concentrate solar heat are:

Photo Above; Courtesy Pixabay.com
Solar reflecting mirrors

Solar Reflectors
- Heliostats
- Fresnel Lenses
- Mirrors
- Solar Troughs, parabolic & fresnel types and Dish
- Solar ponds to trap heat.

These type of devices are collectively termed "Concentrating Solar Power" (CSP) Technologies use reflective mirrors to concentrate sunlight and convert energy into heat. Focused on water in pipes to create steam to drive generators fitted with turbines that generate electricity.

Solar reflectors can be made by using mirrors or by using highly polished metal surfaces that act like mirrors. Mirrors by themselves are great for reflecting solar heat but are unable to track the sun. Heliostats however, are built with tracking motors so that will track the sun maximising the thermal heat collection for the day.

Solar troughs and parabolic dishes are mostly used for the solar heating of water, for hot water or to produce steam for cooking or generating electricity. Both these designs have the characteristic parabolic curved shape, however, one is an elongated trough whilst the other is a disk. Both Solar Troughs and Solar Disks have a reservoir suspended above the surface at an optimal distance to receive solar heat. The reservoir for solar troughs is a metal pipe line running the whole length of the trough. Solar disks normally have a suspended tank containing the heating medium. This heating medium can be:
- Water
- Oil [Heats upto 750°F/393°C]
- Molten Salts [1000°F/538°C]

Water can be used to produce steam directly or solar heated oil or molten salts can be used to boil the water to produce steam.
Concentrated Solar Power [CSP] technology uses three different methods to produce usable energy, these are; Power Tower systems, Trough systems both linear and parabolic as well as

the Dish/engine systems.

The Dish/engine systems are built on a mounting structure utilizing a double axis tracking systems that tracks the sun. Mounted on the reciever is a heat engine. The collected heat is passed through a heat exchange engine of either the Stirling or Brayton cycle engine due to their superior power conversion capabilities.

The Power Tower systems also goes by other names, the most popular is the Central Tower system because the Tower is mostly situated in the centre, surrounded by thousands of heliostats which focus thermal heat to the receiving reservoir that sits at the top of the tower. Most of the more recently built Power Towers use molten salts as the heat transfer medium due to their superior energy storage and heat tranfer abilities. The steam produced is high pressure steam for the electric turbines.

Large CSP systems designed to be used for commercial purposed require large tracts of land to capture the solar radiation to produce electricity on an industrial scale.

When solar heat is focused and concentrated it can reach temperatures in the 1,000's of degrees Celcius. These temperatures can rapidly boil water for steam which is then used to produce motive force, which can turn an electric generator.

More information about CSP can be found on the CSIRO website and also on the ScienceDirect website which explains CSP systems and includes photos and illustrations.

Other excellent photos and illustrations on CSP can be found on ShutterStock.com and GettyImages.com.au which both have a good range of photos and illustrations.

Solar Refrigeration
There are systems that can utilize the heat from the sun to produce cooling, even refrigeration. The following design is one of my own. In my research I came across Steam Jet refrigeration is a proven technology often used on ships. This system can be miniaturised to use on residential homes by using solar steam. The way it works is that you spray a jet of steam across the top of a reservoir of water, which brings most of the heat in the water to the surface rendering the underlying water much colder. Unlike evaporative systems which struggle in high heats and humidity, this system should only get better as it gets hotter, as it is the heat that produces the cooling. You the utilize this cooled water as your refrigerant.

The CSIRO also has Air-Conditioning systems that work on Solar heating. Links to the CSIRO website were mentioned a few pages previously.

Water Purification

Solar thermal can be used to purify water, which can be desalination or it can be disinfecting contaminated water. The main method of disinfecting water these days is the SODIS method. SODIS simply means Solar Disinfection. This method utilises UV from sunlight and infrared to kill any parasites or micro-organisms in the water. Clear plastic bottles are filled with water and placed somewhere like a rooftop to be exposed to maximum sunlight.

Desalination can be achieved by distilling the water by solar methods which leaves almost all contaminants behind. The most common method is to use evaporative distillation. Water that is evaporated into another vessel will be clean for drinking.

Other methods of purifying water include filtration. Water can be filtered at a bulk rate or a little at a time. However filtered water may still need to be disinfected.

Solar Water Harvesting by Condensation

Larger quantities of water can be extracted from the air overnight in what is basically a solar technique. An air well can produce approx 2,000 litres of water per night. This device is like a small swimming pool, or water tank, which has a roof over the top, three lengths of metal pipe are placed through a hole in the top of the roof, giving a tee-pee type appearance. The metal pipes are heated by the sun during the day but at night the cooler air causes water condensation that drips into the water tank. If your water reservoir is large enough, you may want to increase its capacity, this maybe possible through one of the following approaches:

Use additional pipes

Recently found a story about a coating that can be applied to metals that speeds up the condensation process by use of hydrophobic coatings, such as Hydrobead and NeverWet, causing drops to bead together and run off.

Using heliostats or mirrors, to further hear the pipes

See: https://www.sciencealert.com/this-new-material-pulls-clean-drinking-water-straight-out-of-the-air

https://www.scientificamerican.com/article/harvesting-clean-water-from-air/

https://en.wikipedia.org/wiki/Atmospheric_water_generator

Warka Water towers harvest drinkable water from the air
https://www.youtube.com/watch?v=THJVuinPbc0
100 litres per day

Atmospheric tower - potable water generator
https://www.youtube.com/watch?v=ZYaPxbNoO_U
Life expectancy of unit 70 years.
This is a large system

TreeHugger; The Water Seer
Derek Markham; Oct 10, 2016
Wind-powered device can produce 11 gallons per day of clean drinking water from the air.
https://www.treehugger.com/clean-technology/waterseer-can-produce-11-gallons-day-clean-drinking-water-air.html

Solar Stills
Solar stills are another way to collect water without wasting energy, and yet another solar thermal application. A solar still is an eco-friendly way of using the suns energy to purify water. Solar stills make use of thermal energy of the sun to evaporate water and leave behind contaminated residues, thus purifying water whilst UV rays kill pathogens at the same time. In this way solar stills are able to supply clean water for drinking and cooking.

More information on solar stills at:
How Does A Solar Still Work?
Sciencing
https://sciencing.com/solar-still-work-4696783.html
Updated April 24, 2017, By Chris Sherwood

How to Make a Solar Still: The Ultimate Purification Device
In Purification by Jeremiah CasteloAugust 22, 2018

https://worldwaterreserve.com/potable-water/purification/how-to-make-a-solar-still/

How to Make a Solar Still
Mother Earth News
https://www.motherearthnews.com/diy/home/how-to-make-a-solar-still-ze0z1209zsch

Collecting Water from Fog
Another way to collect clean water is to harvest fog. These designs are like nets, sheets of materials that are held up like a wall by posts in the ground. Fog particles condense on the material and flow downwards where they are collected.

One way to make water run off materials faster is to spray them with a new non-toxic sprays called "Hydrobead" or "NeverWet". It is a water repelling, waterproofing in a spray can. One coating is all that you need. This spray would be suitable for metal pipes, material cloth, to speed up water capture for increased capacities of water harvesting systems.

    Hydrobead
    https://www.hydrobead.com/
    NeverWet
    https://www.digitaltrends.com/cool-tech/liquidoff-super-hydrophobic-spray-thats-completely- non-toxic-fabric-friendly/

Author's Evaluation 4

Evaluation of Solar Thermal Systems

Efficiency: Usually takes longer than standard cooking methods.

Cost: Once you have the solar cooking ovens and utensils, it is free.

Usage: You need sunny days, in winter and in overcast conditions may not be worth the effort.

Appeal: Free, non polluting cooking source has to be appealing

Durability: Can be as durable as you make your cooking devices.

Life Expectancy: Depends a lot on durability of materials used in construction.

Limitations: Sunshine is your limiting factor, if you have plenty of sunshine, you're set.

Suitability: Same as for limitations.

Backup: On cloudy or wintery days you may prefer to cook inside.

General Comments: May not be for everyone, but hey, its free energy.

Future: Solar cooking equipment may improve and popularise solar cooking.

Chapter 5

Solar Chimneys

The first solar chimneys were not used to produce electricity but for household cooling.

Solar Chimneys have now come of age, there is now a worldwide market for Solar Chimneys with some of the biggest players including Helioakmi S.A., Solar Innovations Inc., EnviroMission Limited and Specflue Ltd.

The Solar Chimney concept was first used to cool homes and buildings, an almost forgotten technology, that has now re-emerged as a source of artificial wind to produce electricity. This is a proven technology with no emissions.

Solar Chimneys utilize both wind and solar energy to generate electricity. They are large designs that may not be adaptable to households. The idea behind these designs is to create an artificial airflow called a convection current. Air is heated usually at the base of the chimney and rises up through the chimney creating enough pressure to run an electric wind turbine.

These designs have emerged in the last 20 years and are capable of producing upto 20 Mw of electricity per hour. They are a green system, no fuels used, no emissions. Once built the electricity produced is essentially free of cost to the industry. Will these cost savings be passed on to the consumer, probably not.
There are more than one type of Solar Chimney. The main types are:

The Updraft
The Downdraft and
The Crossdraft

The Updraft Model
Most of the publicity regarding Solar Chimneys has been focused upon the Updraft Solar chimneys. The Updraft Solar Chimney has an area around the base which is shrouded, this

area leads to the chimney entrance just about a metre or more above ground level. The Chimney is heated by the sun causing the air inside to heat up and rise into the atmosphere, whilst cooler air from under the shroud moves in to fill the void. This air movement is called a convection current, and can be sustained for approx 20 hours a day, 7 days a week. The air current can be used to drive an electric generator by using a wind turbine. This makes the Solar Chimney much more efficient in electricity production that wind turbines which rely on natural wind.

The Downdraft Model

In a very similar way Downdraft models can be built to harness artificial wind which has been diverted down the Chimney. The generators are usually located at the bottom of the chimney. In order to make a downdraft current water is often employed as a cooling agent. It is finely sprayed into the chimney from the top inlet and descends to the bottom. At the bottom the used water can be collected and reused or allowed to flow into the ocean or other nearby watersource.
Cool air will descend down the tower creating a downdraft. Essentially the idea is the same as the updraft except the airflow is in the opposite direction.

These Downdraft type Solar Chimneys are suitable where there is a watersource. It can be fresh water or it can be seawater, it makes no difference. Some of the designs that I have seen suggest building by the sea and pumping seawater through pipelines and releasing used water back to the sea by pipelines. However if a storage tank were used there would be less pumping required.

The Crossdraft Tunnel

I have not seen any designs for a Crossdraft model as yet, but it appears to me that it has some major advantages over the other two models. The advantages:
Can be built on the ground. Not a plane hazzard
- More stable
- More flexability in design
- Easier to build on the ground.
- Should be cheaper to build, as you can use thin sheet metal in the design.
- Can be built as an array [complete power station]

I have my own designs for a Crossdraft type model. If anyone is interested in this design, make me an offer.

Evaluation of Solar Chimneys

Efficiency: These work on artificial convection currents which gives them the ability to remain active for 20 hours per day in most locations.

Cost: Only for industry unless you own a farm and are willing to invest huge amounts of cash.

Usage: They have been used in the past for household cooling, so it maybe possible to purpose build them into house designs to generate electricity. Recently modified to provide electric current for the grid.

Appeal: Often a personal thing, but to me chimneys never look all that good.

Durability: Good if built well. Could be a problem in earthquake areas.

Life Expectancy: Unknown

Limitations: Most are very high and are an aircraft danger, lights needed to warn aircraft.

Suitability: Only for industry. It maybe possible to produce miniature designs.

Backup: Is possible, batteries, molten salts.

General Comments: Vastly more reliable than wind turbines, can produce large amounts of power.

Future: The power generated is fuel free, clean energy, that makes much more sense than the insanity of wind farms.

Chapter 6

Wind Turbines

Commercial wind generators have sprung up like weeds in recent years. These generators may produce free electricity that does not pollute the air but they are terribly inefficient, and capital cost are enormous for building and installation.

Photo by Peter Gonzales

Wind farms have become popular in many nations. The USA, Australia to name just a few, but they are not only land based, they are even being built in sea locations.

Photo Courtesy https://pixabay.com/images/search/wind%20energy/

Despite their proliferation, wind farms remain inefficient and in the authors opinion a waste of money when better designs already exist. The cost of each of the wind turbines like the ones pictured above in Australian money is $6M AUD ea.
Wind generators operate by converting kinetic energy, the energy of movement into electrical energy The designs for wind energy are adaptable and cost competitive on a large scale. Smaller home systems may work well also but are not likely to meet all of your energy needs.

Despite my dislike for wind generators that rely entirely on wind for them to operate, I still believe that wind power is the most underdeveloped of all the major alternative energy sources.

However, for now, the three main type of wind turbines are as follows:
- VAWT Savonius [Vertical Axis Wind Turbine]
The vertical axis wind turbine is simply a turbine that turns on a vertical axel. There are varying types but the two best designs are the Savonius, and the Darrieus.

Most of the old style windmills were a type of VAWT. The blades of a windmill are equivalent to a turbine. The old Dutch windmills were used to perform all sorts of tasks.

One of the main advantages of VAWT wind generators is that they do not have to be pointed into the wind.

Photo courtesy of Unsplash.com,
by Wim Van't Einde

The Savonius design has an 'S' shaped rotor which can be made with cylinder types shapes such as the old metal rubbish bins, fuel drums or buckets, metal or plastic.
The Darrieus type looks more like the old egg beaters, having three aerofoil type blades attached to a main rotor. A modification of the Darrieus is the "Giromill".
Most VAWTS turn somewhere below 100 RPM

The VAWT designs spin slowly around 100 RPM or less, but have high torque which is best suited to such tasks as; grinding grain, water pumping and other similar tasks but are not efficient in the production of large scale electricity generation, which requires 1000 RPM or above.

- HAWT [Horizontal Axis Wind Turbine]

Once again the explanation is in the Acronym, these turbines spin on a horizontal axis. Most of these designs are considered to supply less power than the VAWT type generators. However they are more suited to areas which experience high wind shear, than the VAWT systems. In high wind shear areas the HAWT generators have an increased of 34% production in electricity. The HAWT type wind generators are better suited to production of electricity than the VAWT design.

In high wind shear locations the HAWT system dominates in production efficiency as the blades always move perpendicular to the wind. HAWT systems are built on high towers and must be very sturdy to withstand high wind shear.

The majority of HAWTS look like windmills with blades that are shaped like propellers which spin on a horizontal axis.

Photo's/pictures of VAWT & HAWT designs See:
     Vertical Axis Wind Turbines - IN 60 SECONDS
     https://www.youtube.com/watch?v=qx_M0nvDIGU

     Wind Power Physics
     https://www.youtube.com/watch?v=qx_M0nvDIGU

     https://www.youtube.com/watch?v=U-50kq7SQVM
     Wind Energy Potential - Sustainable Energy - TU Delft

If you really want a wind generator at home you will need to get good advice about which is the best suited for home installation, and most efficient in low wind conditions.

If you intend to install a home wind generator make you check zoning regulations and ask relevant authorities if there are any restrictions that may cause problems with authorities further down the track.

Commercial Designs
There are a number of wind turbine electrical generating systems available commercially. These turbines can range from hundreds of dollars upto about $30,000- $70,000 for an expensive system. When purchasing wind turbines, keep in mind, you get what you pay for. In other words the cheaper systems may be more like toys. The more expensive units will normally be the more efficient ones.

See: Resource List – BritWind; and Apples to Apples Pt 1

The Nemoi Wind Turbine overcomes several problems of other wind systems, which are big and heavy and require high wind speeds to generate reasonable electrical flows. So the Nemoi system was designed to generate enough power for a four person home at wind speeds of a mere 10-13 miles per hour. This system comes in a package and is easy to assemble, light weight and efficient, compared to other commercially manufactured systems.

The turbines are strong being made from aircraft grade Aluminium and can be installed on the roof of your home.

Claims by Semtive
"Working at very low wind speeds, it is extremely quiet and will not harm the local fauna. Clean wind power generation at a low initial cost with a fast return   on investment. "
[https://semtive.com/residential/]

Semtive also cater for businesses and commercial premises
[https://semtive.com/commercial/]

Change Wind Power
This company claims to have a VAWT type wind generator that outputs 36kW of electricity.
[http://changewindcorp.com/]
See Article:
      Firm After Cheaper Wind Power Wants To Build A Better Turbine.
      BRIAN DOWLING, June 1 /2014
      https://www.courant.com/business/hc-xpm-2014-06-01-hc-change-wind-   turbines-jewett-city- 20140530-story.html

Do It Yourself Wind Generators
There are many designs available, and it can be cheaper to D.I.Y. The hardest part to the D.I.Y. Is getting a decent generator. It seems that the best motors to use were recovered from the old reel-to-reel tape drive computers. These were made by Ametek and were a 99v Dc motor that was used as a generator. These motors are becoming hard to find, although you may still find one on Ebay.

General Notes
Turbines placed in Rural areas have low wind speeds and are not efficient, largely due to wind shear.
Rule of thumb: Small turbine = small power.
To run a house turbines need to be about 25 feet across.
Placing turbines on houses can cause vibrations that can be very annoying, probably best to avoid placing turbine on houses.

See: Resource List

https://www.wired.com/2015/05/future-wind-turbines-no-blades/

Evaluation of Home Wind Systems

Efficiency: Most wind systems are very inefficient due to the fact they they rely upon natural winds, which can blow infrequently depending upon where they are built.

Cost: The cost will vary depending upon weather you decide to purchase your wind generator from a supplier or if you decide to make your own. The price from the supplier can be as high as $35,000 USD.

Usage: Only supply electricity when there is wind blowing. This means that there is plenty of downtime with these designs.

Appeal: Most wind designs are appealing to the eye in general. There are many different designs and variations, so I could not say this about all of them.

Durability: If well built they should last for many years, giving good service when the wind is available.

Life Expectancy: Depending on how well built your wind turbine system is you may get upto 20 -25 years service. Nearly all these designs will have bearings which will eventually wear out and then require replacing.

Limitations: Only work when there is wind.

Suitability: Due to the above factor, I would have to say they are unreliable, unless used in conjunction with artificial winds, which increases their efficiency by an enormous amount..

Backup: They can be connected to a batteryback-up system.

General Comments: I would look for a more reliable system.

Future: There maybe a better future for these designs if someone improves the efficiency of electric generators. Current generators rate approx 18% efficiency under load. They work well, but there is a lot of room for improvement.
The Nemoi wind Generator system could be the future of Home Wind Energy systems if it lives up to its claims.
There is yet huge potential in wind designs that has not been developed. Better designs should yield higher efficiencies.

Energy Storage Devices

Batteries Vary in shapes and sizes to suit todays multitude of Gadgets and Devices. Other storage devices are in development

Chapter 7

Energy Storage – Part 1

Lead Acid Batteries 12v

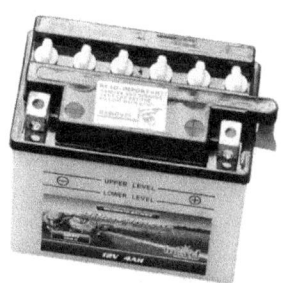 All batteries have some common design components, these include; Casing, Electrodes, and an Electrolyte. The electrolyte is a medium for storing electrical energy. The Electrodes, also called terminals act as your power take off points. There are two type of terminals one is the Cathode and is normally the negative terminal whilst the other is the Anode and is the positive terminal. Electricity flows from electrodes through electrical leads. Electrolytes vary but can include the following types; Acids, Salts, Gels, Polymers, Dry powders and Proton conductors, which are a class of solid electrolytes.

The most common materials used for terminals are; Copper, Zinc, Aluminium, Carbon and Lead.

Connecting Batteries
You can form a network by connecting batteries together in order to increase voltage or amperage or both. There are two types of connections:
- Connecting in Series
- Connecting in Parallel [Linear]

When connecting batteries be sure to use batteries of the same polarity.

To connect in series , the connections need to criss-cross from positive to negative terminals. Using 12 v batteries this connection will give you increased voltage, from 12 to 24 volts.

Connecting in parallel is a linear connection. The negative is connected to the negative and the positive to the positive. The voltage remains at 12v but the Amperage is doubled.

In some of the "Alternative" Lifestyle magazines there have been articles on how to run a household using Car batteries. Most of these homes were in country areas and required modifying appliances to run on a 12V DC sources.

Car Batteries as you know are Lead Acid batteries and they are one of the most common batteries available. These batteries range in price from $150 -$230 AUD new. Second hand batteries may be purchased for as little as $30, however they would need to be tested to ensure that they are still in good working condition.

If setup in a network these batteries can be used as a backup system for Solar generated power or can be used to power a home as the main source of electrical power. This can be done a number of different ways:

- Using 12 V DC or 24 V DC current to run the household lights and appliances.
- Using 12 V DC then converting to 120 V AC or 240 V AC depending on where you live.

The problem with running on Battery power is that the batteries still need to be charged constantly to be used as a consistent reliable power supply. Solar panels can be used to recharge batteries.

Another problem is the lengthy charging times of most Lead Acid batteries. I have heard about quick charging circuits, that can reduce charging times to about 1 hour for a lead acid battery.

Quick charge batteries would be a vast improvement over 12v lead acid batteries that take about 24 hours to fully charge.

Let's have a look at battery types:

- Lead Acid Batteries
- Marine Batteries, designed for boating
- Deep Cycle Batteries
- Dry Cell Batteries
- Lithium Ion Batteries
- Solid State Batteries including Magnesium Core Batteries
- Flow Batteries and MHD Batteries
- Sea Batteries [Mostly for marine applications]
- Molten Salt Batteries [Industrial type batteries, suited as a backup to Solar farms.]

- Super Capacitors and Flywheels are not batteries but are energy storage devices none-the-less.

The list above is not an exhaustive list. It represents the most practicle types of batteries to date.

The most promising of all these batteries are the Lithium Ion batteries and the Solid State Batteries, which are currently undergoing fever pitch research and development by car manufacturers in particular in a race to produce the best battery for the emerging electric car industry. The researchers know that these batteries are not limited to electric vehicles but will have widespread applications including battery backup for solar panels.

Lead acid batteries are still favoured by many people due to their proven performance record, their availability, and the fact that low priced second hand batteries are available from car wreckers, and from battery reconditioners, as these batteries can be reconditioned and reused.

Battery Reconditioning [Lead Acid]
There are websites advocating battery reconditioning to save money on batteries or even use these techniques to turn a profit. See Below.
https://ezbatteryreconditioning.com/

Acid Lead battery restoration recovery 2 of 3 INCLUDES UPDATE 2018
https://www.youtube.com/watch?v=C6p1a_Yw6_Q

How to repair dead dry battery at home , Lead acid battery repairation
https://www.youtube.com/watch?v=KU8AVNI4kZo

Sealed Lead Acid Battery Recovery
https://www.youtube.com/watch?v=2JIzSxqWids

Battery Reconditioning -- All Steps in 1 by Walt Barrett Made in USA
https://www.youtube.com/watch?v=XuZhf2J-NIg

Lead Acid Battery Desulfation Using Epsom Salt --Add Solution to Dead Interstate battery Part 2 of 6
https://www.youtube.com/watch?v=lFEnErs68XM

How to PROPERLY Recover and Recondition a Sulfated Battery
https://www.youtube.com/watch?v=6x-JfckAt20

Evaluation 7

Evaluation of Battery Systems [12V]

Efficiency: Good, these batteries have a proven track record.

Cost: Cheap compared to Lithium Ion batteries, especially second hand.

Usage: Running cars to backup systems

Appeal: Cheap for the moment.

Durability: They are very durable with tough plastic casings.

Life Expectancy: Most are warranted for 2-3 years however they can last much longer before needing reconditioning.

Limitations. They are slow to recharge and do not have the storage capacity of the new lithium ion batteries or the new solid state batteries being developed. Solid State batteries will recharge in minutes according to Toyota.

Suitability: They are suitable for the applications that they are used but Solid State Batteries will outclass them.

Backup: A good source of backup.

General Comments: These batteries will be replaced by some form of Solid State Battery.

Future: Becoming obsolete.

Energy Storage – Part 2

Solid State Batteries

The forerunner to Solid-state batteries is the Lithium ion battery, which are already being used to backup solar energy systems. The main manufacturers are Sonnen and Tesla. Another company that is using Lithium ion batteries for the home solar market is Powervault. More information about Powervault and their batteries can be found on Youtube:

Home Power Storage
Powervault
Fully Charged
https://www.youtube.com/watch?v=8DllPFmXNg

Solid state battery technologies is being driven by the race to market electric cars and Japan is one of the front runners in this race. Japan has a new energy development organisation [NEDO] which is orchestrating the development of new technologies. It has garnered some 23 companies into the race to develop batteries suitable for electric cars. These companies include: Panasonic, Toyota, Nissan, and Honda.

Many other partnerships have sprung up to participate in the development of these new technologies, not just batteries but motors, frameworks, and even materials for the new booming electric vehicle industry.

A relatively new battery type which is emerging onto the marketplace is called a "Solid State Battery". As you can guess these batteries do not rely upon a liquid or polymer electrolyte. This means that the electrolyte is actually a solid. These batteries are a further development of the powdered type like the AAA's.
Research and development efforts have resulted in a battery that has a solid electrolyte that is efficient and quite robust and has a greater energy density than other electrolytes.

The same man that developed the Lithium ion battery is now working on a new ion battery that uses glass and sodium as the electrolyte. The advantage of the glass electrolyte is that it is 3x the power density of Lithium.

The advantages of solid state batteries over other types is that they are smaller, lighter, and greater power density and less likely to explode than any other battery.

These new glass batteries will outperform Lithium ion batteries making them ideal for electric vehicles, possibly increasing the travel range by a factor of 2 or 3 times. This would put them on par with combustion engine vehicles as far as travelling range.

This new type of battery is being eagerly persued by many of the car manufacturers in the race to produce an all electric vehicle that has a travel range similar to that of the combustion engine vehicle.

The main players in this race are:
- Toyota, Honda, & Nissan [Libtec]
- Hyundai [Ionic Materials]
- VW & Quantumscape
- James Dyson and Henry Fisker [E-motion]
- John Goodenough & Maria Brage

Samsung have recently announced that it has tested a new solid state battery based on the lithium ion but has graphene balls in it which allow it to charge 5x faster and enables a 45% increase in capacity which can power an electric car for 600kms.

John Goodenough was the man that originally designed the Lithium ion battery. He has teamed up with Maria Braga who has developed a glass electrolyte. They are currently working on a sodium ion glass battery which promises to outperform Lithium ion by 3x.

VW will be very vigorously researching and developing a solid state battery because it has staked its whole car manufacturing industry on the outcome. It would not surprise me if they find a way to team up with Samsung.

The future of Solid state batteries is looking very solid for now and the race will continue until all these companies are satisfied with their results.

There are more battery types being developed that deserve a mention here, these are:

Gold Nanowire Batteries
Grabat Graphene Batteries
Sulfide superionic battery

At the University of California Irvine Gold Nanowire Batteries are being developed that may result in batteries that don't die. These batteries use Gold nanowires in a gel electrolyte. In recharging tests over three months, having been recharged 200,000 times they have excelled in that there has been no deteriation at all. These are not Solid state as they use a gel electrolyte, however they are very promising none-the-less.

Potentially one of the most superior batteries under development,the Grabat Graphene battery is capable of giving cars a driving range of 500 miles on single charge. The company behind the development of these batteries, Graphenano claims a discharge rate 33 times quicker than Lithium-ion and fully recharging in just a few minutes.

The Superionic battery is being hailed as a "superior battery"which can operate at supercapacitor levels which is ideal for cars especially during acceleration, to offer peak performance.

A consortium of companies is taking part in the development of a Fluoride Ion Battery. These companies are GM, Honda, together with researchers from CalTech, and Nasa's Jet Propulsion Lab in California. Energy density of these batteries is upto 10x that of Lithium ion batteries.

This leaves me to wondering if a battery similar to Samsungs may be 3x more efficient if Sodium were used with graphene balls rather than Lithium.

A new solid state battery is being developed in Tokyo by Tokyo Tech Research by Professor Ryoji Kanno and his team of researchers who were looking to replace Lithium Ion by another safer solid state ion conductor. After much research they came across LGPS Sulfide solid electrolyte which they modified by adding some chlorine to deliver the worlds highest performance ionic conductors. It has slightly higher power density that Lithium ion and higher power output. These batteries were still undergoing development and testing in March of 2018. For more information see:

All-Solid-State-Batteries
Tokyo Tech Research
https://www.youtube.com/watch?v=SiInVYIBVag

Lithium may yet make a comeback with a Lithium hydride solid electrolyte.

The momentum is building towards a major breakthrough in solid-state electrolytes for new batteries that will power electric cars and perhaps a miriad of other devices as well as acting as household storage for solar or wind generated power.

This race will most likely result in a multitude of winners due to the calibre and shear number of entrants in this important historical event.

Here are my concerns and comments about lithium ion, or just lithium in general:
It's Toxic
It a rare earth metal, I'ts not abundant, therefore not sustainable.
As Lithium becomes scarcer it will increase in price, probably dramatically.
It is better for mankind if more common solid ionic electrolytes rise to the top.

Evaluation of Solid State Battery Systems

Efficiency: Will vary according to type
Cost: Will vary according to manufacturers
Usage: From cars and powerwalls, through to emergency supply.
Appeal: Look good
Durability: Tough
Life Expectancy: Check with manufacturers specifications
Limitations: Few
Suitability: Eminently suited to E-Cars, E-Planes, Solar Backup
General Comments: Some are on the market but many still in development stages
Future: Most of these batteries should be avail within 2 years conservatively.

Energy Storage – Part 3

Water Batteries

Another new type of battery that has recently been incorporated into the home energy market is the water battery. This battery stores energy in the form of thermal energy, or heat.

These new batteries will compete with Lithium ion and Solid-state batteries as energy storage devices for residential power supplies.

The big advantage that water batteries have over most other types of energy storage devices is that they can play a dual role, the first is storing excess energy from solar panels for later use, the second is that they also store hot water and can replace the hot water system.

Not all water batteries are designed for this dual role, so if you are looking at purchasing these type of batteries, be sure to mention that you want them for the dual role of hot water system and energy storage.

The manufacturers of Lithium ion batteries are large companies like, Tesla and Sonnen who are large enough to sell their products worldwide.

The Water battery manufacturers have a more limited marketplace, as they are mostly startup businesses which have not been around as long as Sonnen and Tesla. However, the dual role of the water battery gives it a greater appeal than just a storage battery. Neither the Tesla or the Sonnen are designed to give you hot water, it is a bonus that only comes with the water batteries.

More information for water batteries can be found on Youtube, here are Some examples:
1. Sunamp Heat Battery
Fully Charged 12 Oct 2016
https://www.youtube.com/watch?v=9upXeTMHUqE
Stores electricity as heat
runs heating and hot water

After 20,000 cycles with no degredation
Also: http://wattson.energyhive.com/dashboard/AndyT

2. Mixergy Hot Water Tank
Fully Charged 7 Aug 2018
https://www.youtube.com/watch?v=z1Z4JCoPAGc

3. Redflow ZCell batteries for home renewable energy storage
Fully Charged 5th April 2017
https://www.youtube.com/watch?v=4OHstY_kKUY
CEO of Redflow Simon Hackett [Flow Batteries – Very long Life]
Melbourne, Australia.

4. Iron Flow Water Batteries 28 Jun 2018
Large Scale battery – Ideal for Solar farm storage
Battery Fueled by Iron and Water Could Transform the Power Grid
https://www.youtube.com/watch?v=HmtI8Wat7rY
http://www.essinc.com

5. Salt Water Battery
Energy Storage, Clean and Simple
http://aquionenergy.com/
Brand name "Aspen" these batteries are very safe, certified Cradle to Cradle, and are ideal for storing solar energy, they are a deep cycle battery, long lasting batteries. These batteries cannot catch fire or explode. For more information see the website.

6. Edison Battery Construction Nickel Iron
https://www.youtube.com/watch?v=K84PywMwjZg
See here for updated design: http://www.noonco.com/edison/

7. How to Refurbish a Nickle Iron Battery
https://www.youtube.com/watch?v=RxQ-svlFiHU

8. Nickel Iron Edison Battery Install and Rainwater Harvesting System
https://www.youtube.com/watch?v=PIXDv1JI46M

9. Why I Chose Nickel Iron Edison Batteries Over Lithium Ion
https://www.youtube.com/watch?v=2WZ7StJAOuE
Off the Grid Living 16 Feb 2019

10. DIY Projections
How to Make A Powerful Water Flow Battery
Step-By-Step Instructions
https://www.instructables.com/id/How-to-make-a-powerful-water-battery/
By JimJong in Science

Energy Storage – Part 4

Fuel Cells

It has been so long since I've heard anything about fuel cells, I thought that that they had fallen so far behind in development that they were no longer a viable energy storage consideration.

However there is a business gearing up to sell Fuel cells to residential customers. The name of this business is 'Ceres Power' located at Horsham, Sussex. The new fuel cells had to be manufactured from materials cheap enough and common enough to allow for mass production.

These fuel cells use an electric process to convert fuel to electricity, they can use a variety of fuels, including biofuels, but the most convenient would be natural gas, to which many homes already have a connection.

More information is available on Youtube:
> Ceres Power
> Fully Charged, Jun 1, 2017
> https://www.youtube.com/watch?v=PCs9OOHP-0o
> Steel Cell printed ceramics 50% efficient
> Runs on: Natural Gas; Biofuels or Hydrogen
> 1 Kw system
> Factory Location: Horsham Sussex

Ceres Power has designed their fuel cell to be a residential power solution that is wall mounted. The unit looks good on installation and produces power where it is needed. These units produce very little pollution.

This startup business has great potential, but as yet it is only a small operator, if they decide to Licence their design to other manufacturers, it may grow quickly, but if you live outside Britain, realistically, this product may not be available to you.

Hydrogen Fuel Cells

A number of obstacles to the development of Hydrogen Fuel cells have been overcome in the last few years. One of these obstacles was the ability to transport Hydrogen gas in a safe manner, over long distances. This is now possible by using Ammonia as a carrier medium. Ammonia itself is reactive, caustic and toxic.

Hydrogen Fuels cells have been trialled recently in cars that were purposely designed for hydrogen. Essentially these cars are electric cars but use hydrogen fuel cells to generate the electric power source. The cars worked well in the trial and had a range of about 380 miles before refills. This trial revealed a number of drawbacks. These were that Hydrogen refilling stations were only available in the state of california, but electric filling stations are plentiful throughout the entire USA. Is it a good thing to have hydrogen and ammonia together?

The claims that hydrogen can provide longer range and quicker refilling than electric battery powered cars will not last long as the development of newer batteries will soon solve these problems. The designs are already on the drawing board.

See:
How Do Hydrogen Fuel Cell Cars Work?
Drive.com.au
https://www.youtube.com/watch?v=6Ca4s5ZJ0gA

2017 Toyota Mirai Hydrogen Fuel Cell Car Test Drive Video Review
Autobyte
https://www.youtube.com/watch?v=pSHxojRPmnM

Recently in Sandvika in Bærum, Norway the Uno-X hydrogen refilling station blew up, the cause of the explosion is as yet unknown. Raising the issue 'How safe are Hydrogen refilling Stations'?

Comparing the electric battery driven cars to the fuel cell powered cars leads to the conclusion that the battery powered cars have the upper hand with no pollution being produced.

It seems to me that the better choice for fuel cells would be the 'Ceres Power' fuel cells as they can run on biofuels or natural gas. Natural gas is already available at most Fuel service stations worldwide. Whereas Hydrogen fuel powered cars requires specially built storage and filling services.

If "Ceres Power's" claims about the emissions from their power cells being very low, then certainly these fuel cells would be a more practicle solution than Hydrogen fuel cells.

Energy Storage – Part 5

Big Batteries for the Grid

Some developers are now making giant onsite size batteries to be used as storage for the Grid. These batteries are so large that once they are placed onsite they should not be moved. You can get a perspective on the size of these batteries if you have seen a shipping container. Some batteries are housed inside shipping containers.

Don Sadoway believes that by deploying his batteries at power generating stations that they will add an extra 30% power capacity to the existing grid system.

It is obvious now, that every power generating Solar facilities, or wind farm, would do well to employ some type of Big Battery to extend their services throughout the night.

Choice is now available in the Big Battery marketplace with at least 4 manufacturers to select from:
> Don Sadoways Liquid Metal Battery by Ambri
> Giant Lithium Ion Batteries by Tesla
> Iron Flow water battery by Essinc, Ideal for solar farm storage
> https://www.youtube.com/watch?v=HmtI8Wat7rY
> http://www.essinc.com

Salt Water Battery, developed by aquion energy under the brandname Aspen
these batteries are very safe, certified Cradle to Cradle, and are ideal for storing solar energy, they are a deep cycle battery, long lasting batteries. These batteries cannot catch fire or explode.

For more information see the website:
http://aquionenergy.com/

Energy Storage Devices - Part 6

Capacitors & Super Capacitors

A technology that has been showing increasing promise as a possible replacement for batteries are devices called capacitors or as the latest version are being called Super Capacitors due to their rapid increase in storage capacity. In the past capacitors discharged too quickly but the newer Super Capacitors can be discharged much slower almost the same as a battery. The brilliant thing about capacitors is that they can be charged very quickly, whereas batteries are much slower, however this gap is narrowing due to all the research being done for electric car batteries.

Some people are even switching over packs of super capacitors in-place of their car batteries. The super capacitors are much lighter than car batteries and these super capacitor enthusiasts claim that the performance is as good or better than the standard 12 volt lead acid battery. Reference:
https://www.youtube.com/watch?v=gzaLF5tFf88
https://www.youtube.com/watch?v=z3x_kYq3mHM
I would however be cautious of replacing car battery with some of these smaller capacitors as they may not have the durability of a car battery. Power requirements need to be properly matched to the requirements of your vehicle.

Many hybrid and electric buses are now using super capacitors to replace batteries altogether or at least reduce the number of batteries needed. Reference:
https://www.nationalgeographic.com/news/energy/2013/08/130821-supercapacitors/

It seems that capacitors may well replace batteries as they are developing much faster than batteries and are already super quick to recharge, are lighter and smaller, than a battery of similar capacity. They also tend to be cheaper to build. Even the purchase price of similar capacity batteries are much more expensive, giving capacitors an edge in the price department as well. Reference:

https://hackaday.com/2017/01/19/will-supercapacitors-ever-replace-batteries/

Tesla recently purchased an Ultracapacitor manufacturer, which can only be seen as a vote of confidence in capacitor technology. Reference: https://interestingengineering.com/could-ultracapacitors-replace-batteries-in-future-electric-vehicles

It seems clear to the author that capacitors whether you call them Super or Ultra have a very big part to play in future of energy storage.

Chapter 8 – Free Energy

Part 1 Home Generator Systems

Solar Panels
Solar Panels are expensive to purchase but some claim that you can make more efficient solar panels yourself at 85% less cost. Information on how to do this is in E-book form available from Jeff Davis at the DIYKings website. If these panels are as good as claimed, you could arrange them in a 3D array for even more efficiency.

Free or Cheap Electricity

The system that I am about to suggest here has been subject to unjustifiable debunking. The Electricity industry and even the big oil companies have been perpetuating a lie. They claim that you can't run more than one generator from an electric motor to produce usable electricity. They quote "Power in must equal power out" thus claiming that the above idea cannot work. This seems to be a lie.

My rebuttal is this "A body in motion tends to want to stay in motion. In other words once you have something in motion, it takes much less energy to keep it in motion.

Running two generators from a similarly matched electric motor is not a problem. The idea being that you use one generator to power the motor whilst the second generator produces usable power.

Alternators as Motors
A car alternator can be wired to run as a motor.
Tutorial
https://www.instructables.com/id/Car-Alternator-to-Cheap-Motor/

Although it is possible to convert an alternator into an electric motor there is really no need because there are plenty of cheap traction motors from simple DC to more complicated A/C units available in almost any power range you need. They are on Ebay and junkyards all over the USA.

See: Update Free energy forever no wind no solar no gas
https://www.youtube.com/watch?v=TpvXEp1Idis
ThomasBuie EnergyBuies at gmail

Free Electricity
Even if running two generators is out of the question most households only need one. You need a 12v motor to run the alternator, A 12v battery for starting and running the electric motor, and one or two solar panels to charge the 12v battery.

Alternators as Electric Generators
Car alternators can output 90Amps of electricity at 240Volts if wired correctly. Using the power calculation in the back of the book 90Amps x 240 volts = 21.6 Kw, this is enough power to run two standard households. Or it can be used to produce 12V D/C current as required. You MUST use 100Amp fuse between the alternator and to whatever it is being connected.

As solar panels and batteries improve over the next few years, they will be all that you need to fully power your home. You will still need an inverter. Less batteries and less solar panels will be required due to the rapid pace of development.

Free Energy by Spinning Effect
Mark Edwards has developed an electric generator that is more efficient than conventional designs a spinning effect that is said to multiply its power output. You can find this information in his "Power Efficiency Guide" available at DIYeasySolutions.com Mark claims that it is powerful, cheap and easy to build.

Plug-In Power Savers
There are a variety of brands now offering Plug-In Power Savers that can achieve between 30% to 45% savings off your electricity bill. These devices work by balancing out irregularities in your electric current, acting like a power smoother.
A saving of 30-45% is extremely good for a small plug-in device that cost as little as $10 AUD. These devices are easily found on the internet, just do a search on plug-in power savers. I ordered some through Amazon. I am not convinced that these work.

Evaluation of Home Generator Systems

Efficiency: Good
Cost: Free to run
Usage: Electricity production
Appeal: Free is always appealing
Durability: Good
Life Expectancy: Depends on life of Generator which can be replaced.
Limitations: Few
Suitability: Extremely suitable to householder
General Comments: Very flexible. This is a system can be made portable for camping or for a holiday home, caravan.
Future: Free electricity, one less bill.

Free Energy – Part 2

Atmospheric & Earth Electricity

There have been numerous attempts to draw electricity from both the air and from the earth. There has been in most cases limited success. However some methods have been more successful than others.

Some of the successes have been:
- The Tesla Coil
  TESLA POWER Tesla's free energy coil in action!
  https://www.youtube.com/watch?v=VTO9C-OrXys

  How to Make Wireless Energy - Mini Tesla Coil
  https://www.youtube.com/watch?v=12oFWA-bSww
  This small unit is basically a toy but shows you how to transmit power wirelessly.
Larger units would be much more practical, but care must be taken as they are high voltage devices.

Atmospheric Electricity Capture

Tesla Apparatus for Free Cosmic Energy
This device is in very simple terms is a capacitor that is elevated on a stand. The plates of the capacitor are made of aluminium and seperated by a very good dielectric sheet. A framework that is non-conductive to electricity is required for safety and the second plate has power take off wires that direct the current into an inverter for your 120v A/C or 240v A/C supply, depending on our local electrical A/C voltage. The originating current from the device will only be about 12v D/C, so an inverter is needed if you want A/C power.

Nakamats Power Wall
Yoshiro Nakamats, is a very wealthy Japanese inventor with over 3,000 patents. It seems that the Nakamats wall, ie the black antenna, is probably just a very large capacitor. This device powers a house that has 30 bedrooms and then sells excess power to the grid. Both these men

claim that the power comes from cosmic rays. These devices provide electricity 24/7 night and day.

The Nakamats power wall seems to be based on the tesla design with some slight modifications, including the size and colour of the plates.

One of the properties of capacitors is that the larger the surface area of the plates, the more energy is captured. The Nakamats black antenna has a very large surface area, extending along one of the walls of the house.

The reason that aluminium is most often used as plates for this type of device is that Aluminium is a natural antenna material, secondly we know that Tesla used Aluminium sheet in his apparatus.

I don't have the exact specification of these designs, however the calibre of the two main designers speaks for themselves. Details of the Tesla Apparatus can be found at:

NuEnergy
DEVICE TO HARNESS FREE COSMIC ENERGY CLAIMED BY NIKOLA TESLA
https://www.nuenergy.org/nikola-tesla-radiant-energy-system/

TESLA SECRET: ATMOSPHERIC GENERATOR FREE ENERGY
Bedini, Howard Johnson, Gerard Morin, Muammer Yildiz
https://www.youtube.com/watch?v=yHRJvY4tBho
magnetshack, Published on 21 Oct 2015

Free energy antenna
https://www.youtube.com/watch?v=YMO0qvzDMqE
Jimi Hendrix 6/11/2015

Ion Harvesting Technology
The Ion Power Group have demonstrated that they can harvest ions from the surrounding air and produce usable high current electricity that can power lights, electric motors and by electrolysis split water into hydrogen and oxygen or charge Lithium Iron (LiFePO4) batteries and be converted to household current by an inverter.

Whilst the concept of harvesting ions from the air is not new, the process used by Ion Power Group is new as they have patented a carbon nanomaterial to harvest the natural electric charge in the atmosphere in such a way that it produces a clean source of high voltage electricity, both day and night, anywhere on earth.

Ion Power Group
How Ion Harvesting generates Clean Electricity on Earth
https://ionpowergroup.com/how-it-works-on-earth/

Harvesting Atmospheric Electricity!, 28 Feb 2016
The Ion Power Group
https://www.youtube.com/watch?v=fqQGutS2k1Q

It seems that all of the above devices rely upon the arrival from deep space of Galactic Cosmic Rays which when they collide with electrically neutral oxygen and nitrogen atoms produce a partical cascade in the earths atmosphere which results in electrically charged ions in the air both during the day and the night.

Radiant Energy
Tesla's Little Secret
https://www.youtube.com/watch?v=acuc9ARNtfc
Tesla coil "the true secret" how it was really used.
Mr Tesalonian
https://www.youtube.com/watch?v=qwVOp-HPIVE

Earth Energy
It has been known for a very long time that the earth itself acts like a large capacitor, storing extremely large amounts of energy. It is thought that the earth recieves this energy from the sun.

This energy can be tapped and used as a free source of electricity. The ultimate energizer is based upon the work of Nikola Tesla. Information is available at the link below.

Earth Energy
TheUltimateEnergizer.com
Unlimited Free Energy

Currently there is much content about Earth Energy on Youtube.

The Rise of Electric Cars

Chapter 9

Electric Cars

The first cars to hit the roads were fully electric cars developed in the 1830's.
These early electric cars probably used the Nickel-Iron rechargeable batteries. The electric car declined as roads were improved outside urban areas and the electric vehicles could not meet the range that the combustion engine vehicles.

The combustion engine vehicles that pollute the airways still currently on the roads in 2019, these must give way to electric vehicles for zero emissions. Car manufacturers need to ramp up the production of fully electric vehicles. Lets not play around with hybrids.

There are at least some manufacturers who are producing all electric vehicles. One in particular is the Tesla, thanks to Elon Musk.

Photo: Tesla Electric-car
By Richard Shepherd

The time for the electric car is now. In fact it should be the dominant vehicle on the roads by now but governments and car manufacturers are slow to do what they know is right and get behind the proliferation of electric vehicles.

There are already a number of electric cars being marketed worldwide, with most of these relying on lithium ion batteries as their electrical power source. One of the world's leading car makers is about to close their doors on the combustion engine models that they have been making in favour of all electric vehicles.

The move towards producing electric vehicles for the marketplace has meant that a huge amount of research and development is being done not just on developing the perfect battery for an electric vehicle, but the whole drivetrain had to be redesigned to work properly with electric cars.

Research and development continue at a hectic pace as car manufacturers now see the electric vehicle as the near future of automobiles. In the early stages of development much work went into redesigning the framework and panels of the cars so that they were as light and strong as possible. Weight of the car bodies was a concern due to the initial weight of available batteries at the time and the electric motors were of an old design and very heavy.

New electric motors have since been designed to be light weight and have increased performance whilst driving and on acceleration so that they compare favorably with the sporty combustion engines.

More information on new electric motor types for electric cars can be found at:
https://www.greencarreports.com/news/1122363_new-electric-motor-could-eliminate-transmissions

https://newatlas.com/equipmake-electric-spoke-motor-interview/54694/

https://www.popularmechanics.com/cars/car-technology/a25227439/apm-200-eqiupmake-spoke-design/

https://www.youtube.com/watch?v=0Kz_xLeq_Zg

The number ofnew electric motors being developed and emerging onto the market is astounding.

In a bold move, some may call risky, The Volkswagon group made an historic announcement that it will turn its entire production $91 Billion dollars worth, over to electric vehicles. This is good news for the environment.

It's not just the Volkswagon group that are jumping onto the electric car bandwagon but manufacturers worldwide are now taking the electric car market very seriously, partly due to the enthusiasm of the big car manufacturers, and partly due to many design breakthroughs, and the breakthroughs in cost of manufacture.

One of the big expenses in making cars has been in the transmissions, which some of the new designs have been able to eliminate reducing the cost of manufacture by between $8,000 - $10,000 USD. Needless to say this is a huge economic incentive to produce electric cars.

The future of the automotive industry is electric vehicles

The stage has been set for an electric vehicle revolution, the design stages are almost complete, with new higher performance electric motors, and battery development is rapidly catching up to a point where very long range driving is possible.

Electric Vehicle Comparison

Vehicle Make	Range in Miles	Range in Kilometres	Approx Cost in 2019
Kona Electric -A	258	415	$37,495
Chevrolet Bolt A	238	383	$37,495
Nissan Leaf A	150-226	241-363	$30,885
Hyundai Ioniq A	124	199	$31,235
The BMW i3 L	153	246	$45,445
The Tesla Model S L	315-335	506-539	$86,200
The Tesla Model X L	289-295	465-474	$90,700
The Tesla Model 3 L	240-310	386-498	$40,700

Note:
      A = Affordable
      L= Luxury

For more information on these vehicles, see:
https://www.Edmonds.com/Electric-car/articles/best-electric-cars/

Longest range electric cars 2019
https://www.kbb.com/car-reviews-and-news/top-10/longest-range-electric-cars/2100006708/?slide=10

https://electrek.co/2018/12/30/new-electric-vehicles-2019/

Toyota's six new electric vehicles for 2020
https://www.designboom.com/technology/toyota-six-ev-models-plan-06-13-2019/

Over 40 electric cars from which to choose
https://www.cnbc.com/2019/04/23/ny-auto-show-previews-teslas-competition-with-more-than-40-electric-cars.html

Low price electric cars coming soon
https://www.consumerreports.org/hybrids-evs/new-long-range-affordable-electric-cars-coming-soon/
The Electric car revolution has already begun in earnest

Flying Cars 2019

Some people think that flying cars are futuristic but the fact is that flying cars are fact today. Flying cars have been popular as a fantasy theme in science fiction however they are here and now and are being put into production.

Definition
> *A flying car is a type of personal air vehicle or roadside aircraft that provides door-to-door transportation by both ground and air. The term "flying car" is also sometimes used to include hovercars.*
> [Wikipedia, https://en.wikipedia.org/wiki/Flying_car]

These vehicles are not futuristic anymore but they are todays reality. They are not yet mass produced but they are being produced and sold. There are many different designs, some of which will be mentioned below.

The Terrafugia
"The TF-2 is being designed to make travel by air and ground part of nearly everyone's daily commute. It is a three part system that combines a passenger cabin that transfers between a road vehicle and an air vehicle. Passengers travel their full journey, through the air and on the ground without having to switch vehicles along the way.
The TF-2 will take-off like a helicopter, fly like an airplane, and drive on the roads using the latest technology in electric propulsion, construction materials and manufacturing processes to ensure safety and reliability."[https://terrafugia.com/]

Most of these vehicles are very environmentally friendly, they are often called VTOLs, an abbreviation for "vertical takeoff and landing" vehicles.

It seems likely that a large number of these vehicles will be employed in the service industry as Flying Taxis.

The variety of different types of flying cars that are now ready to be manufactured and sold is astounding, some of the designs are:

- Moller M400 skycar, VTOL, on sale on ebay
- Delorean aerospace flying car. Looks like an F1 racing car for the sky.
- Airbus , citiairbus flying taxis.
- Astin Martin, personal air transport in luxury
- Terrafugia purchased by chinese holding company Geely.
- Neva's aquadrone flying quadbike
- Aeromobile has a new flying car that can be pre-ordered.
- Vertical aerospace finished testing UK's first electric flying taxi
- Alaka'i technologies has unveiled a flying taxi that is hydrogen powered

The price of a flying car could be anywhere between $1.6 Million for an AeroMobile flying car to $279,000 for a Terrafugia model.

The next question is of course, how is all this new air traffic going to be regulated. It needs to happen or there would be chaos in the air. Flying insurance is another issue that will need to be put into place.

See also: Electric Bikes, and Flying Bikes.

Evaluation of Electric Cars in General

Efficiency: Becoming competitive with combustion engine vehicles

Cost: The high cost of Electric vehicles is set to drop with the new cheaper motors.

Usage: Range has been a serious issue but new batteries will quench these concerns.

Appeal: Very high appeal due to the low environmental impact. No Pollution

Safety: Should be as safe as any other car on the road. Will have to pass normal testing procedures.

Life Expectancy: Approx 15-20 years but warranties will be much lower.

Limitations: Range, specific to vehicle

Suitability: Extremely suitable as transport vehicle

Comfort: Depends on Brand and make, and personal preferences.

General Comments: These vehicles are a godsend to the environment.

Future: These vehicles along with flying cars are the future, here today

The Nuclear Debate still Rages

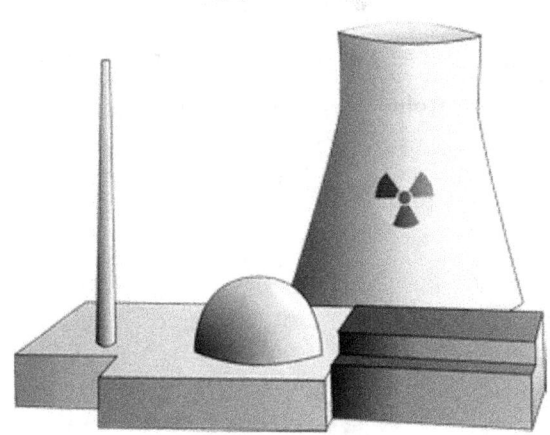

Chapter 10

Nuclear Energy – Part 1

The Old Nuclear Energy

 Current designs are over 50years old and rate a big Fail from me, due to the radioactive waste which has a half life of 25,000 years, there are still activists out there who will tell you that it is clean energy simply because it does not pollute  the atmosphere, most of the time. The dilema has been how do you deal with the radioactive waste.

Nuclear plant operators have been trying to have radioactive waste burried in other nations. Outback South Australia was one chosen location, but as far as I know there was too much public outcry against this notion for it to procede. Would you want radioactive waste burried in your backyard?

Photo courtesy of: Pixabay.com

Photo courtesy: Libreshot.com

Single Cooling Tower
Okostrom meaning: Green Power

4 Cooling Towers of the Temelin
Nuclear Powerstation Czech Republic

In the above photo's we see the familiar sight of a Nuclear smoke stacks. Whilst it is true that the author is against the old nuclear power plants, there are many with opposing views. In 2019 the Nuclear Energy debate still in full flight.

The Case Against Nuclear Energy
The opposing arguments against nuclear power are very strong and include the following:
High level radioactive waste from spent nuclear fuel.

- Accidents
- The link to Nuclear weapons, and nuclear weapons proliferation.
- National security
- Cancer Risk
- Limited Sites
- Costs
- Competition with Renewables

High Level Radioactive Waste

The very long term radioactivity of wastes generated by nuclear plants is a very real concern. Even now in 2019 there are still no answers to the long tern storage of nuclear waste materials.

Each nuclear power station produces radioactive waste which must be multiplied by the 444 nuclear power stations in 30 countries. It all adds up to massive amounts of nuclear waste which nobody wants, and there are no solutions to its containment.
The facilities that have been built as storage for nuclear wastes are running out of space leaving the nuclear industry to resort to costly and less safe options.

The combined world's nuclear powered fleet creates high level nuclear waste to the order of 10,000 metric tons per year. Management of these highly radioactive materials are a dilema facing world governments. Some of these high level radioactive wastes include:
- Technetium-99 [half life of 220,000 years]
- iodine-129 [half life of 15.7 million years]

- neptunium-237 [half life 2 million years]
- plutonium-239 [half life of 24,000 years]

The good news is that some of these wastes can be used again as fuel, extracting energy from these waste products is called nuclear recycling.

Accidents

Although physists will tell you that modern nuclear power plants are safe due to built in safety features, the fact still remains that accidents can happen. Radioactive materials being vented into the atmosphere is just one of the potential dangers of nuclear accidents.

These have been a number of serious nuclear accidents around the world since nuclear power plants were introduced. Thes accidents include:
- Chernobyl explosion in The Russian Ukraine.
- Mayak nuclear weapons production plant near Chelyabinsk in Russia
- The Fukushima disaster
- Three Mile Island accident

Chernobyl

This was reported as a steam explosion which happened on 26 April 1986 at the Chernobyl Nuclear Power Station situated in the Ukraine. The explosion resulted in radioactive contamination spewing into the atmosphere. Clouds of this toxic contamination was carried over much of Western USSR and Europe on the prevailing winds.

Containment of this catastrophy required the utilization of over half a million workers and cost the Soviet economy a crippling 18 billion rubles.

Chelyabinsk

These accidents occurred in 1949, 1957 and 1967 and together rated higher than the Chernobyl explosion in total radioactive emissions. Luckily for this region it was much less populated than Chernobyl and surrounds.

The Fukushima Disaster

In Japan on 11 March 2011 immediately after an earthquake and Tsunami, the cooling system at Fukushima Daiichi nuclear power plant failed. The resulting explosions and fire caused breaches in the system that released dangerous levels of radiation.

Radiation leaked into the air, soil and water and has been called the worst nuclear incident in 25 years, with 50,000 householders forced to move.

The disaster sparked a meltdown in the Japanese stock market, as well as panic buying in supermarkets.

The 3 Mile Island Accident

The Three Mile Island incident happened in Middletown Pennsylvania on 1 April 1979. It is the worst nuclear incident in the history of the United States. It was a core meltdown caused by the failure of a pressurised water reactor at the Three Mile Island Nuclear Generating Plant.

The Link to Nuclear Weapons

There is a huge concern that using and developing nuclear energy power programs may increase the widescale manufacture of nuclear weapons. The same nuclear fuel used for power production can just as easily be turned to weapons manufacture. There is great risk if the nuclear fuel and technologies fall into the wrong hands. Countries that have unstable governments and or high levels of corruption should be discouraged from becoming nuclear nations.

National Security

Due to the fact that nuclear explosions are so devastating, Nuclear power plants are an enticing target for terrorists. Such an attack could put major population centres at risk due to the radioactive material that would be emitted into the atmosphere and subject to prevailing winds could cover a huge area possibly even thousands of kilometres.

Cancer Risk

Following nuclear fallout there is a hightened risk of cancer for those living near nuclear sites, such as power plants, especially at risk are children. Leukemia in children is much higher in those living near nuclear power plants that that of the general population. Nuclear power workers are exposed to higher rates of radiation than normal, and as a result cancer rates in these workers is similar to the survivors of the Japanese nuclear bombs.

Limited Sites

It is important for nuclear reactors to be close by a source of water for cooling and these sites need to be drought free, hurricane free, and earthquake free to prevent the triggering of a nuclear disaster. The number of sites that meet these requirements is very limited. Climate change does not sit well with these requirements.

Costs

Whilst renewable energy costs are going down, the costs associated with nuclear energy are on the rise. Many of the current plants are on the verge of being shut down due to cost factors. The Fuel costs, capital costs and even maintenance are far higher for nuclear plants.

Competition with renewables

Investment and development of nuclear plants, security, mining infrastructure and other associated costs all draw away limited investment available for cleaner sources such as solar, geothermal and wind generating plants. Funding for electric power generation is already limited, and increasing nuclear capacity will only increase competition that will detract from funding for renewables.

The Case for Nuclear Energy

Supporters of nuclear energy claim that nuclear power is clean and sustainable, reliable source of energy. They point to the amounts of uninterupted energy that does not pollute or emit carbon emissions that add to the cause of global warming. It is further suggested that nuclear energy provides plentiful electric power, well paying employment, energy security, and a reduced dependence on imported fuels. All good arguments.
The thought is put forward by supporters that nuclear power produces almost zero air pollution and compares it with the dirty carbon emitting power generating plants such as those that burn coal oil or natural gas.
Proponents also state that nuclear energy is the only way to meet the 'Paris Agreement' to reduce carbon emissions.

Bloomberg seems to think that Nuclear Energy should stand with the other renewable energies such as wind, solar and geothermal.

The 'Boston Globe' even wrote an article claiming that nuclear plants were important in the fight against global emissions and limiting climate change.
It seems that there is now a growing support for nuclear energy coming from high placed officials such as retired admirals and generals, even national security officials. They have all tried to persuade Rick Perry 'Energy Secretary' to ensure that nuclear power stations remain

prominent and valued in the US electricity markets, and remain recognised by policy makers to ensure national security.

Evaluation of The Old Nuclear Systems

Efficiency: Highly Efficient

Cost: Capital costs very high

Usage: The production of electricity

Appeal: Many people are opposed to nuclear due to potential for deadly radioactive emissions

Durability: Non-Issue

Life Expectancy: 30-40 years is the maximum expected lifespan.

Limitations: Non-issue

Suitability: This has to be measured against the potential for disasterous explosions.

Backup: Non-issue

General Comments: Cannot be considered clean energy as nuclear plants produce radioactive waste
with half life of 50,000 years. Proper containment systems have never been designed, current containment systems for storage of radioactive waste are only considered to be temporary measures.

Future: These systems should be taken offline and dismantled as they pose too high a risk not just for the hosting nation but if radioactive material is ejected into the atmosphere, strong winds could spread over it for thousands of kilometres, over neighbouring nations. This is an unacceptable risk.

The Future of the new Nuclear industry looks bright. New designs by Taylor Wilson have secured the Nuclear industry a place in the green energy movement. This young genius has redesigned Nuclear power plants so that they are much safer, produce no radioactive waste, and are cheaper to build than the old plants. I believe Taylor Wilson also said that they can use radioactive waste as fuel in the new reactors. He said they eat it up.

Nuclear Energy – Part 2

Taylor Wilson

To understand the current Nuclear thinking it is necessary to go back to the end of the 70's and early 80's in Germany where massive unrest and upheaval of the vast majority of the populace had mass rallies against Nuclear Energy. This decision to become a nuclear free nation came about during this time of enormous anger amongst the populace against nuclear energy. The move away from nuclear was said to be a step towards a low-carbon economy.

This seemed to be contradictory to many people outside the germanic states, especially due to the fact that germany replaced most of its nuclear stations with coal fired stations. In 2013 Germany had reduced its nuclear capacity to 13% from a high of 29.5% in the year 2000. It is further set to reduce its nuclear capacity to zero by the year 2022. This phase-out was called "Energiewende" in German, meaning energy transition.

Protests against Nuclear energy had been ongoing since the 1970's.

The accident in the US Nuclear power plant at the Three Mile Island site in 1979 sparked massive protests in Germany adding to a growing fear of nuclear technology.
Many protests were also centred around locations of radioactive wastes, processing and storage sites.

The anti-nuclear movement in the germanic states was fueled by genuine fear and a degree of panic which was heightened by the Fukushima disaster in 2011 and the Chernobyl catastrophe in April of 1986.

An opinion poll conducted in March 2015 revealed that 81% of the German population was in favour of the governments decision to phase out nuclear powerstations.

Germany was not the only nation to turn its back on nuclear energy. Other nations comitted to phasing out nuclear energy include Italy, Belgium, and Switzerland. Nations that are comitted to remaining nuclear free include, Denmark, Ireland, Portugal and Austria.

Currently 31 countries operate nuclear power plants. China is one of these nations but its nuclear capacity remains at 10.5% of overall electricity production.

However things seem to be about to change in the nuclear arena with the advent of a totally new nuclear system about to be put into place. In early 2019 it was announced by Taylor Wilson that he had a new design that was safe and free of radioactive waste material. Genuinely Clean Nuclear energy was something that I never thought that I would see in my lifetime, but if Taylor Wilson gets the financial backing that he needs, then nuclear energy will become popular as a source of clean energy.

Until very recently I was totally against Nuclear Power due to the radioactive wastes produced by the reactors. A problem that our scientists have not been able to solve. One young man however, Taylor Wilson has designed nuclear stations that are safe and do not produce the radioactive wastes like the old nuclear plants.

This young man Taylor Wilson is a true genius. He is now just at the age where he could be starting University [College]. Amazingly he saw the flaws in the old nuclear reactor designs and he thought about these flaws and how to overcome them. He found a way. He has designed his own Nuclear Plant, that eliminates all the problems associated with the old [current] designs.

Taylor Wilson's design will produce clean energy without the radioactive wastes. This is a truly amazing achievement and deserves high praise.

This is the type of Nuclear Power that I could support wholeheartedly and is on par with Molten Salt Thorium Nuclear plants.

It's too early to evaluate this new nuclear technology but it seems very promising.

Nuclear Energy – Part 3

Thorium Reactors

The initial concept of the Molten Salt reactor was first revealed in 1950 as part of an early aircraft reactor experiment. It was not acted upon until the early 1960's when the US Government commissioned the building and operation of a Thorium molten salt nuclear fission reactor at The Oakridge National Laboratories site, at Oakridge in the USA, to determine the viability of the concept.

The Fuel used was Thorium and the coolant used was Lithium Fluoride. This was a Lithium Fluoride Thorium Reactor [LFTR], which was sometimes called a Lifter.
The program at Oakridge was very successful, easily proving the viability of this type of reactor.

The reactor built at Oakridge was a 7.4Mw test reactor. It was operational for a period of approx. 4 years, and reached operational temperature of 650°C. It was built with walk away safety measures. In essence this means that if a problem occurred and the reactor began to overheat, a plug made from lithium salt would melt and all the fuel and coolant would be drained into holding tanks where they would cool to normal temperatures. This feature eliminates the possibility of any type of explosion.
This wonderful safety feature ensures Thorium reactors are safe to operate.

The Thorium Molten Salt reactor program was only a minor program that was being developed concurrently with the Fast Breeder program which was considered a major project in comparison. The interest in this type of reactor waned with the emergence of the Fast Breeder reactors. Funding for the program was cut in 1976 as the Atomic Energy Commission redirected funds into the Fast Breeder program which had already received large amounts of Government funding.

Other nations also built Molten salt reactors but many strayed from using safe fuels and coolant systems. The United Kingdom and Russia both built their own versions. The Russian program ground to a halt after the Chernobyl incident.

The Nobel prize winner 'Alvin Weinberg' has recently thrown his support behind the Thorium Molten Salt Reactors, as the fusion reactors have not yet become a viable alternative. Problems with these Fast Breeders has not yet been resolved.

Thorium Molten Salt Reactors on the otherhand are a proven viable safe option. The radioactive material waste byproduct can be recycled as fuel again, and then be burnt in a radioactive waste burner, which was developed inline with the Molten Salt Reactors.

Why has this technology been hidden away so long? The answer to this question is probably political, however it is time to bring this technology into the limelight. Thorium cannot easily be used to make nuclear weapons, this maybe one reason it has been overlooked.

It is now time to forget about making nuclear weapons and focus on producing clean energy. Thorium Molten Salt Reactors have a huge potential for generating large amounts of electricity at very cheap prices.

The old nuclear plants based on uranium fuels should be decommissioned and dismantled and then replaced with Thorium Molten Salt Reactors, which will provide clean, cheap, safe energy into the future.

Thorium as a fuel has an extraordinary energy density in that, an amount the size of 3 m&ms could produce enough electricity to last you a lifetime. It is extremely abundant in the soil, in every country, and has low radioactivity in comparison to uranium.

Due to the fact that these Reactors can be made smaller means that they could be modified to use in spacecraft, possibly even submarines and ships of the oceans fleets.

If they can be made small enough, they may be used as a powersource for skyscrapers and tall towers.

Evaluation of Thorium Nuclear Energy Systems
Efficiency: Very high efficiency
Cost: Less than the old designs, can be built much smaller than other Nuclear plants. Running costs are very low as Thorium is very abundant worldwide, unlike uranium which is expensive because it is scarce.
Usage: Production of electricity, highly efficient.
Appeal: Clean Energy is highly appealing.
Durability: Non-issue
Life Expectancy: TBA
Limitations: Should be imposed on both the Fuel and Coolants used. Fuel should remain as Thorium and Coolant should remain as Lithium Fluoride. Experimentation is no longer required.
Suitability: Extremely suitable for producing electricity for the grid. Power source for spacecraft.
Scaleability: Can be made smaller, for spacecraft, nuclear fleet, or larger for large scale generation.
General Comments: Great design with walk away safety features built into reactor plant. Easily on par with the new Taylor Wilson wilson designs for fission reactors. It may even have a slight edge over the Wilson Taylor designs due to the cheapness of the Thorium fuel.
Future: Clean Nuclear Energy has a very bright future. There is also a nuclear waste burner to eliminate radioactive wastes.

Nuclear Energy – Part 4

Nuclear Fusion

Nuclear fusion is seen as the holy grail of electrical energy generation due to the fact that it works by combining two atoms rather than splitting them as in the fission reaction, which caused radioactive waste, nuclear fusion does not result in dangerous radioactive byproducts. It is a clean reaction that produces massive amounts of electrical energy.

There are now many nations trying to develop Nuclear Fusion Reactors for the production of electrical energy. Just to name a few of these nations, are; China, USA, France and also Australia.

One of the main problems facing the development of Fusion Reactors is the extremely high temperatures that are required for the reactions to occur. It also means that the materials used in the reactors must be able to contain these enormous temperatures.

China has been working for decades trying to build a reactor that can withstand temperatures in the millions of degrees in order to contain a fusion reaction. The Chinese thus far have been unsuccessful in their attempts to build a safe working fusion reactor.

France and Australia have been working together in France to develop fusion reactor technology.

Although it is known that the USA is also working on Nuclear Fusion technology there has been little encouraging news come the American efforts.

Australian Hydrogen Boron Fusion Reactor
by Loz Blain Feb 21, 2020
zpenergy.com

An Australian company with the odd name of HB11 has been working on the development of fusion reaction technology and has recently made the following claims; It has produced a working fusion reactor that does use or not produce any radioactive materials and that it does not require super high temperatures to operate.

Its Australian company director says that its Hydrogen Boron Fusion Reactor technology is working much better than expected.

Nuclear Fusion Reactors may become a common feature of the electricity generation industry sooner than any of us expected.

Chapter 11

Emergency Power Supplies

Emergency power supplies may be needed under a range of different circumstances. One of the most common situations in which you may need to use an emergency power supply is in a blackout. If you know that it will only be for a short duration you maybe able to wait it out.

However if the interruption to supply is going to be for an extended period of time you will need to have a backup electrical supply ready to utilize. In very cold areas, when you lose your electricity supply it could also mean that you lose heating as well. This could be a critical loss, potentially even lethal. Although, a wood fire stove or heater, or fireplace with a chimney, would be expected in this type of climate.

Many modern dwellings may have become so heavily reliant on the electricity grid that they may not have a backup source of power. This could be a critical mistake, in the case of severe storms, especially if your home is off the beaten track.

If you live in areas that can become snow bound, be sure to have a backup power supply of that is suitable to your needs.

If you have a gas heater that will last until the gas runs out then you are back to having no heat.

There are a number of options, which maybe lifesaving, these include having a battery backup, solar, or even generators. The problem with petrol powered generators, besides the noise, is the fact that they will run out of fuel unless you have a large supply of petrol on hand.

A battery system, that is where the batteries supply the main power but would need to be recharged, which can be done using solar panels, this is free electricity. Or it can be done the other way around, with Solar as the main source and batteries as back-up, or night source of power.

A solar emergency system would need battery backup, but once again its free electricity.

A 2x generator system that is run by an electric motor should run for absolute years until either bearings wear out or the belt wears out, but these can be replaced and the system can be up and running again. The most logical design is to have a battery to start the system, once started turn off the battery, as the system should be self running from that point onwards. A 12v motor should be used, so that it can be started by the battery. One generator should be 12v in order to keep the motor running, and the other generator can be A/C [120/240v] depending on your needs and how you set things up.

This system can be small enough to be portable, although it could be a little heavy. Can be mounted on a trolley for ease of movement. This would make it ideal and easily mobile for powering a campsite.

Back-up Systems you can Purchase

You could also invest in an Uninterupted Power Supply [UPS] or a back-up power supply such as the Kodiac Solar Generator.
https://learn.eartheasy.com/articles/emergency-power-supply-are-you-ready-for-the-next-major-power-outage/

See LifeWire for the 7 best UPS
https://www.lifewire.com/best-uninterrupted-power-supplies-4142625

D.I.Y. Build A Power Back-up System
Instructables
https://www.instructables.com/id/How-to-Build-a-Battery-Backup-Generator/

How to Build Your Own Uninterruptible Power Supply: 13 Steps
https://www.wikihow.com/Build-Your-Own-Uninterruptible-Power-Supply
Jul 14, 2018
This one simply produces AC power with a continuous duty inverter

See also: Chapter 8 Part 2 Atmospheric & Earth Electricity
The Ultimate Energizer

Chapter 12

Tidal Electric Energy Generators

In earlier times in both North America and Europe tidal dams were built. They would fill with seawater on the high tides turning waterwheels to mill grain, these mills date back to the middle ages in Rome.

A source of electric power that is sometimes overlooked is Tidal Electric Generating Systems. The worlds first Tidal Electric Power Generating plant was built in 1966 at Rance in France. It remained the world's largest Tidal Power Generating system until 2011 when South Korea built its lake Sihwa Tidal Power Generating Station as part of a sea wall barrier, complete with 10 turbines set into the seawall generating 254Mw of electricity.

More Tidal Electricity Generating plants around have been built worldwide mostly in the last 15-20 years. There are many different designs but the 4 main type of Tidal generating plants are one of the following basic types:

- Tidal Stream Generators
- Tidal Barrage
- Dynamic Tidal Power
- Tidal Lagoon

Higher efficiencies can be achieved with Tidal generators than wind turbines as water has a greater density than air. In terms of power, using the same size turbine, an ocean tidal current of 10mph or 8.6 knots has an energy equivalent of a 90 mph wind speed.

These Tidal Stream generators are often fixed to shallow sea floors and work the same way as wind turbines, having rotors under the water-level, that are turned by the tides of the sea as they ebb and flow.

Tidal Barrages are constructed as a low dam wall, known as a barrage, often used as part of a protective seawall which creates a tidal lake or reservoir behind it. Barrages can stretch from a few hundred metres to kilometres in length. Underwater tunnels are cut into its width and fitted with sea gates to control the flow of water. Fitted inside these tunnels are a number of turbines, as the water gushes past the turbines tidal electricity is generated. Electricity is only generated when the water is flowing, when the water is still during high or low tides no electricity is being generated.

Dynamic Tidal Power is as yet an untested idea. The concept has been designed but there are as yet no DTP type generating systems have been constructed. Until one is built and tested they remain a promising technology.

Tidal Lagoons generating systems require the construction of circular walls with generators built into, embeded into the walls, creating reservoirs and these could be built in double or triple formations. It is expected that this type of configuration will deliver very efficient tidal energy. Turbines can be built to operate on both incoming and outgoing flows.

For many years there has been talk of building a Tidal Energy station at the inlet Bay of Fundy. A small scale version was eventually built in 1984 with a 20 Mw capacity.

There is another system that works on temperature difference of ocean waters which is called OTEC. It relies on ammonia as a working fluid in a sealed vessel which turns to gas at approx 32°C, the gas is used to turn a turbine thus generating electricity. After going through the turbine the gas condenses back to liquid ammonia. This way the working fluid is continually recycled and reused.

Australia is another candidate for Tidal Power Stations with its enormous coastlines. Under an Australian Government initiative Australia's tidal energy resources have been mapped out by Australian Renewable Energy Agency [ARENA]. The study that was done by the government was to unlock the potential for Tidal Power generation in Australia.

Tidal Power developers now have this feasability study to work from and be able to build projects where they will have the least environmental impact and are viable sites.

Already there are a number of Tidal energy sites in Australia, the largest being in at

Gladstone Ports in Queensland where electricity is already flowing into the grid. Battery storage will be used for the time between tides where power is not being generated.

The potential of Tidal Energy to supply the worlds energy needs is enormous and is similar to the potential of solar energy which can supply the worlds need for energy many times over.

Wikipedia has a list of current Tidal energy projects and those under construction, which shows that harnessing energy from the waves is becoming more prevalent. Tidal energy is a very clean way to harness energy, no pollution, no carbon, just clean consistent energy on a daily basis.

There are a few minor drawbacks that are concerned with environmental issues and on shipping and navigating waters that contain Tidal generators, but these issues pale into insignificance when compared to global warming.

Pinterest has over 200 Photo's of Tidal Energy Generators.

Evaluation of Tidal Energy Systems

Efficiency: Very high efficiency
Cost: Minimal as no fuel required.
Usage: Production of electricity, highly efficient.
Appeal: Clean Energy is highly appealing.
Durability: Non-issue
Life Expectancy: Anywhere between 20 years to 120 years depending on maintenance and corrosion.
Limitations: Limited to Industry
Suitability: Extremely suitable for producing electricity for the grid.
Scaleability: Possible to scale down to household supply but issues of ownership of beach and zoning would arise.
General Comments: Unless you own an Island, or live in a very remote location, leave it to industry.
Future: Tidal is truely a clean energy source and being run by the tides there is an inexhaustible supply waiting to be tapped.

Chapter 13

Industrial Air Purifiers

Greenhouse Gas Capture and Extraction

According to World Health Organisation reports 9 out of 10 people are breathing polluted air with over 4.2 million people are dying every year from this exposure.

Air filtration is not the answer in itself but is another step towards cleaner air. The use of industrial air purifiers is a very positive move towards cleaner healthier air.

If your city centres suffer from heavily polluted air then it make sense to install some air purifiers in the city centre. The deployment of the air filters should be where they are most needed.

There are a number of different types of industrial air purifiers, some of these are:
World's Largest Air Purifier Helps Reduce Haze in Northwest China City
Beijing China
https://www.youtube.com/watch?v=vQioJnH1e1A

Purevento Gmbh
https://www.purevento.com/en/purevento-city-air-cleaner/
Germany

City Tree
This Moss-Covered Air Purifier Can Do the Work of 275 Urban Trees
Kathleen Villaluz, August 13th, 2017
https://interestingengineering.com/moss-covered-air-purifier-can-work-of-275-urban-trees

Ceramic honeycomb air filters could cut city pollution

Catherine Collins, OCTOBER 30, 2018
Horizon: The EU Research & Innovation Magazine
https://phys.org/news/2018-10-ceramic-honeycomb-air-filters-city.html

Teen serial inventor returns with pollution filter to clear city skies
By Zeena Saifi, Daryl Brown and Tom Page, CNN
March 28, 2018
https://edition.cnn.com/2018/03/28/health/angad-daryani-tomorrows-hero/index.html

China's giant air filter system is solar assisted, not requiring any other power source.
India will follow suite inspired by China's Xian city air cleaning towers. Kurin Systems will build 40 foot towers in Delhi that can clean the air for a radius of 2 miles, each tower will be capable of cleaning 1,130 cubic metres of air per day.
These new towers have been nicknamed the 'City Cleaner'.

For more information See:
Boss Magazine
Are Giant air purifiers the solution to polluted cities
https://thebossmagazine.com/air-purification-towers

Other locations where the most air pollution will occur is in the industrial zoned areas of each nation. It may be prudent to have air purifiers every square kilometre depending upon pollution measurements, in these industrial zones. The EPA in each city should be able to work out the requirements.

City tree is vertical trellis type structure with moss and lichen attached that act as a natural outdoor air filter, having its own water supply and solar panels, only requires a few hours maintenance per year.
The city Tree concept is innovative and compact, but you would need a lot of them spread out through the city, however they are cheap, small, efficient and natural. They filter 240 tons of carbon dioxide annually, and also filter nitrogen oxides, ozone, and particulate matter.

In Paris a new type of city outdoor honeycomb airfilter has been schedule to be completed by 2024 to debut at the Paris Olympics. This new model has been awarded The Horizon Prize of

3 Million euro for its innovative design.

The working mechanism of the filter is the honeycomb design as air is sucked through the structure any particulate matter in the flow adheres to the walls and only clean air is emitted. The filter is made from a ceramic material known as Cordierite, which contains a mix of clay, alumina, talc and silica.

Daryani a 19 year old from Mumbai left school in the nineth grade and gained employment with MIT Media Laboratories. After taking part in many start ups and colaborating on a string of inventions, he decided to move to the USA and manufacture industrial scale City air filters to combat airbourne pollution and carcinogens hang like clouds of gloom above cities.

Air quality has become a major international health issue affecting all the major population centres of the earth, although India and China have suffered some of the worst pollution. The result is health issues such as lung diseases, like Asthma, and even cancers.

Technologies are also developing to extract CO_2 from the air and convert it to Ethanol, or fuel products like diesel

The fact that CO_2 can be extracted from the air is very good news, but there is a lot to extract.

How One Company Pulls Carbon From The Air, Aiming To Avert A Climate Catastrophe
https://www.npr.org/2018/12/10/673742751/how-1-company-pulls-carbon-from-the-air-aiming-to-avert-a-climate-catastrophe

A Canadian Company called Carbon Engineering operating out of Squamish, British Columbia are working on a pilot plant that is designed to pull carbon out of the air in a process known as "direct air capture". The result is white beads or pellets that look like bean bag stuffing. They are able to turn these white pellets into a clear synthetic fuel.

Two other companies doing the same thing are Climeworks and ExxonMobil and Global Thermostat who in agreement with Carbon Engineering's cost evaluation.

Climeworks is a Swiss company that is extracting carbon from the air through a recently unveiled Direct Air Capture plant, and is supplying that CO_2 to a nearby greenhouse to fertilise tomatoes and cucumbers.

The process of pulling carbon from the air is not really cost effective as yet. But at the cost of less than $100 per ton, its a process that should be encouraged, if we are to win the fight against CO_2 in the air, but the best way is to stop putting it there in the first place. Reducing the amount that is going into the air together with extracting huge amounts from the air should see CO_2 levels begin to return to normal natural levels.

It may be better to recycle this carbon as fuel rather than refining more natural oil which is very polluting just in the refining stages.

Carbon Engineering has already designed a plant to extract 1 Million tons of CO_2 from the air every 12 months. Several more of these units would be needed. At least two each of these units need to be operating in China and India.

6 Ways to Remove Carbon Pollution from the Sky
https://www.wri.org/blog/2018/09/6-ways-remove-carbon-pollution-sky
- The six ways that we can remove carbon from the air is "Plant more Trees" as they naturally remove CO_2. Restore existing forests and create new forests, plant more trees even in cities.
- Restoring carbon in farm soils, carbon is good for soil health.
- Biomass, must be carefully managed to have a positive rather than a negative impact.
- Direct Air Capture, already discussed above.
- Seawater Capture, by extracting CO_2 from the oceans, these waters then draw in more CO_2 from the air to regain balanced levels.
- Enhanced Weathering, as weathering occurs CO_2 bonds with different minerals, and so ends up being held in minerals and soils.

Recent scientific breakthroughs in technology have resulted in the ability to extract CO_2 from the atmosphere and transform it back into coal. This is good news, that CO_2 can now be stored as coal.
More Information at:
> Scientists Just Pulled CO_2 From Air And Turned It Into Coal
> https://www.forbes.com/sites/trevornace/2019/02/27/scientists-just-pulled-co2-from-air-and-turned-it-into-coal/#3fb129ae4563

Trevor Nace

Exxon Mobile and Global Thermostat are gearing up to make huge profits from extracting CO_2 from the atmosphere and selling it for reuse to multiple large and growing industries. CO_2 can be used to manufacture many other products, including plastics and oils.

Extracting Methane from the Air
It is now possible to capture and extract methane from the air, this new technique should diffuse the concern about atmospheric methane. Some sources would lead us to believe that atmospheric methane is a major problem. This sort of talk is not in line with the known facts, and is exaggerated and deceptive. It is however reassuring to know that this methane can be captured and removed.

Innovation: Methane capture gives more bang for the buck
NewScientist:
By Kate Ravilious
https://www.newscientist.com/article/dn18977-innovation-methane-capture-gives-more-bang-for-the-buck/

Scientists Discover New Ways to Capture Methane
April 18, 2013 by Kirsten Korosec
https://www.environmentalleader.com/2013/04/scientists-discover-new-ways-to-capture-methane

"It's been a banner week for hydrocarbons made from waste gases. Earlier this week, a company announced that it had delivered 4,000 gallons of jet fuel made from steel-plant waste gases to Virgin Atlantic. Now, Swiss company Climeworks has announced the opening of a new plant in Italy that will collect carbon dioxide (CO_2) from ambient air and pair it with renewably made hydrogen (H_2) to make methane fuel that would add little or no CO_2 to the atmosphere." [Company that sucks CO_2 from air announces a new methane-producing plant, Megan Geuss, 10/5/2018]
https://arstechnica.com/science/2018/10/company-that-sucks-co2-from-air-announces-a-new-methane-producing-plant/

Capturing Methane from the air and liquidizing it for fuel may be a better alternative environmentally speaking.

Four of the best methods and materials for the capture of greenhouse gases ranging from Coffee grounds to Ion Exchange Resins can be found in the following article:

> Four Cool Technologies That Clear the Air – The Future of Carbon Capture
> By Jeremy Deaton July 22, 2016
> https://www.popsci.com/four-cool-ways-to-clear-air/

A Bio-engineer Frederiek-Maarten Kerchhof from Ghent University in Flanders has discovered bacteria that can remove methane from the air in a more cost effective method that is more efficient than any other current method, and these bacteria can collaborate with other types of bacteria that can remove polluting substances from groundwater.

> Ghent scientist uses bacteria to remove methane from air
> FlandersToday
> http://www.flanderstoday.eu/innovation/ghent-scientist-uses-bacteria-remove-methane-air
> by Andy Furniere, journalist Saturday 13 July 2019

The number of Methods and materials being developed to capture and extract greenhouse gases from the atmosphere are growing rapidly as a direct result of international concern for the possible environmental impact of such gases on the health and wellbeing of people and the flaura and fauna of the earth.

Chapter 14

The Resurgence of Oil and Gas

It is now certain that Oil and Gas will make a comeback challenge to fuel the worlds energy markets. In the midst of the clean energy frenzy, who would have thought that the Petrochemical industry, would be on the verge of a comeback?

The pollution problem with the combustion of these fossil fuels has not been addressed, the price of fuels has been at ridiculous levels for decades. The world has been led to believe that there was a shortage of fossil fuels and under this guise the price of fuels has been hijacked to unprecedented levels. In short fuel prices have been an insult to the consumer.

It is becoming apparent that fossil fuels are here stay as vast shale oil reserves are now able to produce cheap oil due to new extraction methods, much of it in the form of shale gas. America has huge supplies of shale gas, and Brazil has massive offshore oil reserves, whilst northern Alberta has discovered huge oil sand deposits.

These recently discovered oil fields will be capable of supplying oil to the world for the next century at current consumption rates.

Fossil fuels will compete with clean energy sources well into this century and possibly beyond.

The worlds reliance on fossil fuels cannot be switched off at a moments notice but should be scaled back at an achievable rate, one which can be as fair as possible to all the energy providers.

Politicians need to work with the energy providers not against them. Regulation of the Energy sector by politicians could be a very large step backwards mainly due to their unreasonable expectations. In particular the Paris Accord needs to be reviewed or thrown out in favour of a workable solution, that is not going to bankrupt either the energy providers or governments as they transition to cleaner energy.

The politically driven paranoia about climate change and global warming needs to be toned way down due to lack of any significant changes, or evidence. Politicians need to listen to genuine professionals in the field of atmospheric science, and climate such as Nils-Axel Morner and Hadi Dowlatabadi, and Richard Lindzen.

Summary Page

Household Free Electricity systems: [Free to run]
    Solar Panels with Battery Backup
    Fresnel Lens to Drive Steam engine
    Car Alternator, DC Motor, 12v Battery with enough solar panels to recharge the battery.
    Batteries as main power with Solar rechargeing

Industrial Free Electric Systems [Free or almost free to run]
    Solar Farms/Solar Parks
    Solar Thermal
    Wind systems
    Tidal
    Geothermal

Free to run City Air Purifiers
    Solar Powered Air Purifier [Xian China]
    The CityTree [Germany]

Free Water Purification
    SODIS [Solar Disinfection], scale is normally household level.

Evaluating Global Warming

I don't like pollution in any form and I believe that mankind needs to be good caretakers of this planet, as it's the only one we have and every generation should share this same responsibility.

Statements that were coming from the press and from the scientific community, and world leaders almost convinced me that global warming was caused by mankind, which encapsulated global warming in the following terms:

The problem of atmospheric pollution is a particularly daunting one when considering the threats posed by manmade greenhouse gases which trap extra heat in the atmosphere leading to global warming, and climate change. The cost in lives has already been staggering with China reporting over a million deaths per year because of air pollution.

Acid rain caused by air pollution can kill trees and devastate forests.

This problem of air pollution has been building since the industrial revolution with the rise of factories, the invention of steam trains which burned coal to heat the boilers, and later the invention of the internal combustion engine for motor cars, all contributed to the air pollution problem we face today. There had been very little interest in fixing air pollution until recent years, when scientists discovered that air pollution is a major cause of Climate change, and global warming.

Since World leaders and scientists became aware of the problem, they have expressed concern and seem to be actively trying to address the situation, or are they?

What has become obvious is that there is not just one simple solution but all the major causes of air pollution need to be addressed at the source. These major causes are:
- Factories, that have smokestacks.
- Automobiles that run on fossil fuels
- Powerstations that run on fossil fuels, especially coal.

Governments may need to impose stricter guidelines on industry and ensure that these guidelines are being adhered to by sending out inspectors who can asses and report back to the government. In most countries when it comes to air pollution most factory are self regulating. Governments should be more involved in air pollution control initiated by industries to reduce air pollution to a very minimal level. I am sure that most responsible governments set a standard for air pollution within their industrial sectors, but are their industries being held accountable?

The Automobile industry is certainly on the right track with most manufacturers now developing their own lines of electric vehicles. Electric vehicles have recently been redesigned entirely. Battery designs keep improving, as car designers can now see a great future for electric cars, and the better the battery the better the perception of the electric car by the consumer, which will translate into greater sales down the line.

Power supplies for residential homes and cities is becoming much greener with the growth of solar farms. There are also companies making huge storage batteries for solar farms which will make electric power available 24/7. This is a major breakthrough because the highest power usage is at night time when the sun is not shining above.

Many residences are now opting to install solar cells with a battery backup system so that excess electricity can be used at night time. This is a great way to produce clean energy.

Some Energy systems are definitely better than others, but the system best for you will depend on:
- the cost to set it up
- existing systems
- your own preferences

However, the overall best system for households depends on the needs of the person or family. It is best if it can be made to be portable for easy mobility. It needs works in all weather conditions. Once it is built there should no other costs except for maintenance as needed. It is should be adaptable, for a variety of uses i.e.:

- Electric Cars.

- Household Electricity.
- Camping.
- Backup power.
- Holiday homes.
- Caravans.

The Ultimate Energizer seems to tick most of these boxes and without emissions.
I still believe that clean technologies can only be good for mankind and for the planet. I fully support the move to cleaner technologies.
However, arguments against man being the cause of global warming throw a huge shadow over the opposing view.

Mankind produces only 3% of atmospheric carbon dioxide with nature producing the other 97% of carbon dioxide in the air.

In the light of the very small contribution that mankind has made to greenhouse gases the arguments that suggesting mankind is responsible for global warming are simply ridiculus. The fact that politicians in high places are supporting mankinds culpability in global warming is very suspicious when these same political forces are extracting huge amounts of money from governments worldwide.
When experts in the field like Nils-Axel Morner, a former head of the Paleo-Geophysics and Geodynamics Department in Stockholm are being ignored and hushed, suspicion must grow, smacks of a very big cover-up.

Global warming itself is a fact but Anthropogenic global warming is a scam. The true cause of global warming is the cycles of the sun which do have a very big impact upon climate, even more so than the North Atlantic current.
Today's warmer solar cycle is expected to revert to a cooling solar cycle soon.
It is strange that very few people are even considering the sun as the cause of global warming. The cycles of the sun and how they affect climate is a very interesting topic and one that more people should educate themselves about, to avoid being scammed in a similar way in the future.

"In the year 2000 Al Gore said that the Florida keys would be underwater by 2016...... It's February 19th 2019 and I'm sitting here at the same Beach I've been going to for 20 years in

key West Florida and the level of the ocean hasn't risen at all"..... [Joe Lavine]

JULY 10: *"Germany and Netherlands have frost which is affecting their farming. Learn the truth, the planet is not experiencing global warming. We are entering a grand solar minimum".* [Andrew Hill]

Global warming happened between 1978 – 1998 and reach a height of an extra 4 tenths of a degree celcius. Then it stopped, despite greenhouse gases rising slightly.
We know this is true because of Satelite temperature records. Upto 40% of Climate forcasts and computer models that incorporating global warming have been proven wrong, failing to match current climate facts. With the march of time more and more climate models are failing.
Antarctic ice has grown significantly, but the Climate alarmists have refused to acknowledge that fact.

The impact of global warming was built up to be a huge catastrophic event whereas in reality the effect on the earth and on mankind have been minimal.

Alarmists groups include the IPCC and NASA which are exaggerating climate changes, for unknown reasons, do they must have a secret agenda? They have made false claims that 97% of scientists believe in manmade climate change. This could have simply been a mistake on their part. The real figure is more like 30% and falling, as climate model after climate model fail and even Satelite temperature records have show a pause in the warming.

The slight variations in temperature that have occurred are within normal variations and not a cause for concern. The earth has cycles and patterns of climate that have been churning for extremely long periods of time, climate patterns that we have yet to begin to understand.

Nils-Axel Morner, has pointed out that Carbon Dioxide is actually good for the earth and particularly for plants. He even went as far as stating that the earth could use more CO_2.

The earth has coped with CO_2 levels in the past that were much higher than they are today. In rainforests under the canopy CO_2 levels of upto 6% have been recorded, and yet you would struggle to find a healthier environment. Remember that in most other places the CO_2 level is about 0.04% of the atmosphere.

You can easily deduce from this that we have not yet reached a dangerous level of CO_2 and its not likely to be reached, or caused by man in the near future.

There is now very little evidence to suggest that global warming was anthropogenic, that is manmade.

There is more reason to be concerned about nature which produces 97% of CO_2 in the atmosphere than with man who only releases 3% of total CO_2 into the air.

Topic List of Resources

Air pollution & Global-warming
Air Pollution Everything You Need to Know
Jillian Mackenzie November 01, 2016

https://www.nrdc.org/stories/air-pollution-everything-you-need-know

Accessed: 12/05/2019

Effects of global warming
https://www.livescience.com/37057-global-warming-effects.html
By Alina Bradford and Stephanie Pappas
August 12, 2017 09:12am ET

7 Ways global warming is affecting daily life
Environmental Defence Fund
https://www.edf.org/card/7-ways-global-warming-affecting-daily-life

"Decreases in global beer supply due to extreme drought and heat" [PDF], Nature Plants

"The world's food supply is made insecure by climate change," United Nations

National Climate Assessment

"Great Barrier Reef: Only 7% not bleached, survey finds," ABC News (Australia)

"Resilience potential of the Ethiopian coffee sector under climate change" [PDF], Nature Plants

"A Brewing Storm: The Climate Change Risks to Coffee" [PDF], The Climate Institute

Batteries
    Lead Acid

Lithium
Solid State Batteries

Electric Cars – See References

Motors & Generators
The Newman Motor [Big Eureka]
Joseph Newman
https://www.youtube.com/watch?v=VOiCBJMj-FM

The Searl Effect Generator
by John Searl
https://www.youtube.com/watch?

Nuclear Energy- See References
Thorium
Why making energy from dirt might save the world | Rusty Towell | TEDxACU
https://www.youtube.com/watch?v=jDqCpfVwdP4

Making Safe Nuclear Power from Thorium | Thomas Jam Pedersen | TEDxCopenhagen
https://www.youtube.com/watch?v=tHO1ebNxhVl

Solar Energy
A "Eureka" for Solar Energy | Bert Conings | TEDxUHasselt
https://www.youtube.com/watch?v=wkCmX24PRKE

Storing solar energy in the strangest places: Will Chueh at TEDxStanford
https://www.youtube.com/watch?v=aFaOr05gvpl

A new chapter in energy storage | Danielle Fong | TEDxDanubia
https://www.youtube.com/watch?v=-QNwGuqh9O0

Solar Chimneys
Solar Cooking
Cantina West

www.solarcooker-at-cantinawest.com/solarcooking-howitworks.html
Solar Panels
Solar Thermal CSP

SuperCapacitors

Wind Turbines /Wind Power

Wind Turbines & Generators
Http://www.mdpub.com/wind_Turbine/

https://www.instructables.com/id/How-I-built-an-electricity-producing-wind-turbine/

https://homesteading.com/diy-wind-turbine-generators-living-off-the-grid

http://theselfsufficientliving.com/diy-wind-turbine-designs-to-generate-off-the-grid-power/
https://greenterrafirma.com/DIY_wind_turbine.html   [Tutorial]
The above contains underscores

https://www.goodairgeeks.com/best-diy-wind-turbine-ideas/

The Nemoi Wind Turbine

The following links from Instructables are for building the Turbine only, as they are not connected to a generator.

VAWT [D.I.Y]
https://www.instructables.com/id/Building-a-Vertical-Axis-Wind-Turbine-VAWT-/

VAWT Savonius [D.I.Y]
https://www.instructables.com/id/Build-your-own-Savonius-VAWT-Vertical-Axis-Wind-T/

Motors for use as Generators for Wind systems
Http://www.tigwindpower.com/ametek.htm

Wind Project
http://www.velacreations.com

Wind Turbine Projects
http://www.otherpower.com

Wind Turbine
BritWind
by Fully Charged
https://www.youtube.com/watch?v=pcdXLbzGVpo

Wind turbines [good power at low rpm]
Apples to Apples Part 1
https://www.youtube.com/watch?v=CTZi2dyFdKk

References

Global Warming & Climate Change

Global Warming
Causes and Effects of Climate Change
Https://www.nationalgeographic.com/environment/global-warming/global-warming-effects/

Global Climate Change
Https://climate.nasa.gov/effects

The Arguement against manmade global warming

New sun-driven cooling period of Earth 'not far off'
https://www.youtube.com/watch?v=ViY2J3LPgN4
Nils-Axel Morner, a former head of the Paleo-Geophysics and Geodynamics Department in Stockholm, says a new solar-driven cooling period for the Earth is 'not far off'.

Sky News Australia
Published on 18 Jun 2019
Global Warming Hoax, Misleading Humanity, False Information – United   Nations
Sea levels not rising, 1mm per year overseas, no rise at all in Australia, stable for 70 years.

The Grand Solar Minimum
https://www.youtube.com/watch?v=YgjjmnpYTqQ
Tinyhouse Prepper
Published on 11 Jun 2019
Evidence for the coming Grand Solar Minimum and the Mini Ice Age. For more information, you can watch

WHY HAS THE SUN GONE TO SLEEP? - BBC NEWS
https://www.youtube.com/watch?v=DueVWamHmYs
Subscribe to BBC News www.youtube.com/bbcnews Scientists are saying that the Sun is in a

phase of "solar lull" - meaning that it has fallen asleep - and it is baffling them. So what's going on? Rebecca Morelle reports for BBC Newsnight

The Next Ice Age May Be Sooner Then You Think
https://www.youtube.com/watch?v=_GQo_1W0lww
This winter was pretty bad - but imagine having to live through a harsh winter all year long. Fortunately - the Gulf Stream stops that from happening - but   what if there was no Gulf Stream? More in tonight's Daily Take.
The Big Picture RT
Published on 2 Apr 2015

A Funny Thing Happened on the Way to Global Warming
https://www.youtube.com/watch?v=RZlICdawHRA
Steven F. Hayward, Pepperdine University This lecture is part of Hillsdale College's 12 Nov 2014 CCA series. To learn more about Hillsdale College and the CCA programs, visit
http://www.hillsdale.edu/outreach/cca

Climate I: Is The Debate Over?
https://www.youtube.com/watch?v=gJwayalLpYY
Richard Lindzen MIT Prof. Atmospheric Sciences [Climate Change Skeptic]
Hadi Dowlatabadi, Uni British Columbia
The Agenda with Steve Paikin
Published on 11 Mar 2010

Solar Cells & Panels

Solar Cells & Panels
https://seia.org/initiatives/photovoltaics

Solar Energy Technologies [Photovoltaics]
SEIA [Solar Energy Industries Association]
WWW.seia.org
April 2018

Singularity Hub:
https://singularityhub.com/2019/05/17/5-coming-breakthroughs-in-energy-and-transportation

Solar Cooking

Solar Cooking In India
Solar Thermal Magazine
https://solarthermalmagazine.com/

Solar Battery Backup

"Where are Tesla's factories based? Including Elon Musk's gigantic gigafactory vision in China."
By Felix Todd.
https://www.compelo.com/tesla-factories-elon-musk-gigafactory

In Australia for tips on Solar go to: lp1.solartips.com.au

Redflow ZCell batteries for home renewable energy storage | Fully Charged
https://www.youtube.com/watch?v=4OHstY_kKUY
Flow Batteries: As Storage for Solar Cells

Batteries Part 1 – Lead Acid Batteries

Lead Acid Battery
Wikipedia
https://simple.wikipedia.org/wiki/Lead_acid_battery

Sealed Lead Acid Battery
Wikipedia
https://simple.wikipedia.org/wiki/Sealed_lead_acid_battery

Valve Regulated Lead Acid Batteries [VRLA]includes Gel Batteries
Wikipedia
https://en.wikipedia.org/wiki/VRLA_battery

Deep-cycle Batteries
Wikipedia
https://en.wikipedia.org/wiki/Deep-cycle_battery

Lead Acid Battery: Working, Construction and Charging/Discharging
BySourav Gupta, Dec 19, 2018
https://circuitdigest.com/tutorial/lead-acid-battery-working-construction-and-charging-discharging

Battery Reconditioning
There are websites advocating battery reconditioning to save money on batteries or even use these techniques to turn a profit. See Below.
https://ezbatteryreconditioning.com/

Batteries Part 2-- Solid State Batteried

Solid State Batteries
Https://www.marketecheasier.com/solid-state-batteries

Solid state Battery
Wikipedia
https://en.wikipedia.org/wiki/Solid-state_battery

Toyota, Nissan, Honda, Panasonic partner to develop solid-state batteries
Posted June 21, 2018 by Charles Morris & filed under Newswire, The Tech.
https://chargedevs.com/newswire/toyota-nissan-honda-panasonic-partner-to-develop-solid-state-batteries/

Honda-presents-new-battery-chemistry-that-could-succeed-lithium-ion
Eric C Evarts: 21 December 2018
https://www.greencarreports.com/news/1120563_honda-presents-new-battery-chemistry-that-could-succeed-lithium-ion

Future Batteries, Coming Soon
Max Langridge and Luke Edwards
Subjects: Gold Nanowires, Grabat Graphene Batteries, Sulfide Superionic battery
https://www.pocket-lint.com/gadgets/news/130380-future-batteries-coming-soon-charge-in-seconds-last-months-and-power-over-the-air

Wind Turbines

Wind Turbines
https://www.survivalrenewableenergy.com/top10-10-best-home-wind-turbines

Wind Powers
https://greenterrafirma.com/wind%20turbines.html

202: Types of wind turbines and their Advantages and Disadvantages
Kohilo University
http://kohilowind.com/kohilo-university/202-types-of-wind-turbines-their-advantages-disadvantages/

Electric Vehicles

Electric Vehicles
Edmonds
Http://www.edmunds.com/electric-cars/articles/best-electric-cars/
By Will Kaufman April 16th 2019

Electric vehicle
Https://en.wikipedia.org/wiki/Electric_vehicle

Greenhouse Gases

Greenhouse Effect
Https://www.environment.gov.au/climate-change/climate-science-data/climate-science/greenhouse-effect

Greenhouse Gas
Wikipedia
https://en.wikipedia.org/wiki/greenhouse_gas

What Are Greenhouse Gases
David Suzuki Foundation
https://davidsuzuki.org/what-you-can-do/greenhouse-gases

Greenhouse Gases – Energy Explained, Your Guide to Understanding
https://www.eia.gov/energyexplained/index.php?page=environment_about_ghg

Nuclear Energy

Green America Magazine
10 Reasons to Oppose Nuclear Energy
https://www.greenamerica.org/fight-dirty-energy/amazon-build-cleaner-cloud/10-reasons-oppose-nuclear-energy

Wikipedia
https://en.wikipedia.org/wiki/Nuclear_power_debate

The Nuclear Debate
http://www.world-nuclear.org/information-library/current-and-future-generation/the-nuclear-debate.aspx
Accessed: Friday 10th May 2019

Nuclear Energy Debate: The Climate is Changing
By Matt Wald December 20, 2018
https://www.nei.org/news/2018/the-nuclear-energy-debate-the-climate-is-changing

The History behind Germany's nuclear phase-out
Kerstine Appun, 02/1/2018
https://www.cleanenergywire/factsheets/history-behind-germanys-nuclear-phase-out

Taylor Wilson

Taylor Wilson: My radical plan for small nuclear fission reactors
https://www.youtube.com/watch?v=5HL1BEC024g

The Future of Energy | Taylor Wilson | TEDxUniversityofNevada
https://www.youtube.com/watch?v=bB93E-C7OPo

Nuclear Thorium

Molten Salt Reactor
https://en.wikipedia.org/wiki/Molten_salt_reactor

Renewable-energy Sources

Renewable energy
Wikipedia
https://en.wikipedia.org/wiki/Renewable_energy

Student Energy
Renewable-energy
[https://www.studentenergy.org/topics/renewable-energy]

Biomass
Energy Explained, Your Guide to understanding Energy
EIA [Jun21, 2018]
http://www.eia.gov/energyexplained/?page=biomass_home

Regenerative Coal
Advanced coal technologies improve emissions and efficiency
Kennedy Maize 1/11/2018
https://www.powermag.com/advanced-coal-technologies-improve-emissions-and-efficiency-2/

Tidal Energy Sources

Tidal power trial shows promise for new wave of renewable energy developments
Laura Gentry 23 Nov 2018
https://www.abc.net.au/news/2018-11-23/tidal-power-new-wave-renewable-energy-development-queensland/10544862

Unlocking the Potential of Australia's Tidal Energy
https://arena.gov.au/news/unlocking-potential-australias-tidal-energy

Australia-Led tidal energy project sets new production records

Giles Parkinson
https://reneweconomy.com.au/australia-led-tidal-energy-project-sets-new-production-records-29054/

Tidal Power
Wikipedia
https://en.wikipedia.org/wiki/Tidal_power

Alternative Energy Tutorials
Tidal Energy Generation
http://www.alternative-energy-tutorials.com/tidal-energy/tidal-energy.html

The Resurgence of Fossil Fuels

Actually, Fossil Fuels Are Here to Stay
The Atlantic Magazine
CHRYSTIA FREELAND
JULY/AUGUST 2012 ISSUE
https://www.theatlantic.com/magazine/archive/2012/07/actually-fossil-fuels-are-here-to-stay/309039/

Why Fossil Fuels Survive
The Washington Post; 28/2/2018
By Robert J. Samuelson
https://www.washingtonpost.com/opinions/why-fossil-fuels-survive/2018/02/28/

Power Calculations

Watts = Amps x Volts

The Above equation is the most important power calculation with which you should need to work.

Helpful sites for power conversions:

Metric Power Conversion
    Casio
    http://keisan.casio.com/exec/system/1305519671

    Rapid Tables
    https://www.rapidtables.com/convert/power/index.html

Horsepower to Wattage
    Conversion
    https://www.rapidtables.com/convert/power/hp-to-watt.html
This site also converts Watts to Horsepower.

The End

Dedication

This book is dedicated to all those who care for the environment and
dislike all kinds of pollution. It appears now that
mankinds ability to pollute the planet and its atmosphere
maybe resulting in potentially devastating changes in
climate that could affect every man, woman and child on earth.

To those of you who are promoting green technology and clean energy
sources, I wish to thank you and request that you continue on that
course. You are the heros of this present age.

Alternative green energy sources which do not emit any kind of
pollution will GREATLY help to reduce atmospheric gas emissions.
My own designs are all geared to be environmentally friendly,
and funds from this book will be used to prototype these designs.

By purchasing this book you are not only helping me,
but also the environment, and the air that we breathe.

About The Author

Hi There, I'm Richard Shepherd. Like most of you, I hate air pollution. Personally I have a passion for Technology and Invention. This book was written with a view to raising awareness of the looming climate change crisis that is now engulfing the planet. It is an attempt to combat the problem at its source. The greatest assets in this fight against air pollution and climate change is you. Clean technology is the tool that we need to defeat old dirty power systems, by having them replaced with eco-friendly power solutions.

I am an inventor but have lacked the funds to develop my own clean energy devices, so I am writing this book, to help you in the fight against global warming, and to raise funds to develop my prototypes which are all geared to be eco-friendly.

I am asking you to buy my book "Clean Energy Sources Beyond 2020" together we can achieve a mutual goal. Together we can help save the planet.

Other Publications by this Author

Clean Energy E-Book Series

1 Alarming News on Global Warming

2 Empower Yourself with Solar Energy

3 Awesome Wind and Tidal Energy

4 The Extraordinary Battery Revolution

5 Amazing Guide to Free Home Energy

6 Imminent Invasion of Incredible Electric Cars

7 The Mind Blowing Nuclear Debate

8 Secret Energy Sources Catalogue

Paperback

"The Birth of Opal"

Qualifying Statement

When I say that you can have Free electricity or Zero cost electricity, I am talking about free running costs. Let's be very clear about this. The equipment needed to provide that electricity will come at a cost, unless you already have it sitting around at home somewhere.

If you choose the right system you should be able to minimise that cost also.

Cost between different systems will vary greatly, but in the end there is a choice to be made, and it is my purpose to make that choice a little easier for you by providing you with alternatives and then comparing them for you by my evaluations.

Once you have your system set-up, then the running cost can be free or cheaper depending on the type of equipment that you install.

At some stage you will probably have to maintain or even replace some components as they may have worn out from continuous use.

Remember that most things manufactured these days will only last 15-20 years due to built in obsolescence. However, 15 – 20 years of free or cheap electricity must be a blessing, and you can save an enormous amount of money over this time.

I am not going to tell you that any system will run forever without wearing out or breaking down

If you maintain your equipment, then you could have enormous savings on electrical costs for the rest of your life.

Disclaimer

The content provided in these pages is for information purposes only. The author will not be held responsible for any loss or damage, resulting from the use or misuse of this information whether it be by accident or intention, if you act upon it you do so at your own risk. Safety remains the responsibility of the reader.

Electrical equipment can be lethal. If you are unsure of what you are doing, seek the help of a qualified electrition.

The author is not responsible for the accuracy of third party information which has provided for your convenience.

It is your responsibility to be aware of local laws, and zoning regulations. The author encourages readers to respect and keep to the law.